ACTA
BIBLIOTHECAE R.
UNIVERSITATIS UPSALIENSIS
VOL. XXXI

A History of Non-Printed Science

A SELECT CATALOGUE OF

The Waller Collection

BY

MARCO BERETTA

ACTA UNIVERSITATIS UPSALIENSIS

This catalogue is also included in the series of
UPPSALA UNIVERSITETSBIBLIOTEKS UTSTÄLLNINGSKATALOGER (No. 32)

&

UPPSALA STUDIES IN HISTORY OF SCIENCE (No. 14)
Office for History of Science
UPPSALA UNIVERSITY

Published with the aid of grants from
MARIANNE & MARCUS WALLENBERGS STIFTELSE
KABI PHARMACIA AB

Jacket illustration: Portrait of Paracelsus, see the Catalogue, No. 7

The catalogue text is set on a Macintosh IIsi, typeface Bembo
Printed on book paper Linné 120 g, from Lessebo Paper Mill
Layout and production: Xtina Wootz

ISSN 0346-7465 ISSN 0502-7462 ISSN 0282-1036
ISBN 91-554-3070-8

Printed in Sweden by MO PRINT, Uppsala 1993

TO THE MEMORY OF
ERIK WALLER

Fig. I. *Erik Waller*.

PREFACE

In 1950 Erik Waller decided to donate to Uppsala University Library his collection of books dealing with the history of medicine and science. Some three year later the Library succeeded in purchasing Waller's collections of letters, manuscripts, Alba amicorum, medals, and engravings. As a whole this was the largest acquisition made in the history of this prestigious library. To give an idea of the magnitude of this acquisition it suffices to mention that the Waller Collection attained in quantity the number of volumes acquired by the University Library in its first 164 years of existence (1621-1785).[1] This, however, does not do full justice to a private library which in fact is in many ways unique in the world.

The richness of Waller's collection of books is largely known from the catalogue published by Hans Sallander in 1955 on the basis of notes left by Waller.[2] Evidence of its success is provided by the recent publication of a new edition in the United States.[3] In contrast to the publicity and the fame surrounding the Waller Collection of books, the collection of manuscripts, letters, medals, etc., is hardly known to the international community of historians of science. This is serious because this part of the collection is even more interesting than the book collection, and contains many unique and extremely rare items. Furthermore, its quantity, consisting of more than 35,000 items, offers to historians an unexplored treasure-trove of manuscript sources of scientific thought dating from 1255 up to the first half of the 20th century. The primary reason for the oblivion into which this part of the Waller Collection has fallen, is the lack of a catalogue illustrating its contents. This lack is not surprising because such an achievement would require far more expertise, financial support, and time, than the cataloguing of Waller's books required. However, in the hope that sooner or later a general catalogue of the Waller collection of autographs and manuscripts will be undertaken,

---

[1] On the development of the holdings of Uppsala University Library, see the essay by Claes Annerstedt, "Förteckning å Upsala universitetsbiblioteks ledare jämte korta upplysningar rörande deras verksamhet", in *Uppsala Universitets Biblioteks Minnesskrift 1621-1921,* (Uppsala, 1921), pp. 1-21.

[2] Hans Sallander, *Bibliotheca Walleriana. The Books illustrating the History of Medicine and Science collected by Erik Waller and bequeathed to the Library of the Royal University of Uppsala,* 2 vols., (Uppsala, 1955).

[3] A facsimile edition in 400 copies of *Bibliotheca Walleriana* was published by an American bookdealer (New York, 1991).

I thought it would have been of some utility to select some valuable items from the collection and provide them with a preliminary description. In the course of this work, I made a preliminary survey of the whole Collection, the provisonal results of which are set out in the second part of the Introduction.

When I finished selecting the 70 items included in this catalogue, I was in doubt whether it would be more appropriate to provide them with a thorough and accurate description as required in scholarly critical editions of manuscript sources, or to give just a few bibliographical references such as author, title, date, dimension, and shelfmark. Encouraged by the help I received from many friends and scholars, I decided to make a compromise, that is to say, to make some historical investigation into each item and to transcribe the most significant of them. The immediate and primary purpose of such an approach was to provide historians of science with some tangible evidence of the scientific importance of the Collection. The transcriptions of a few letters and the historical annotations to them will hopefully, even if they do not meet all scholarly needs, inspire new research on a collection which for nearly 40 years has remained unexplored. For this purpose, a bibliographical list of names and works, prestigious as they might be, would not have served as well.

The selection presented in this catalogue follows a chronological order, though not too strictly. I have tried to select items which reflect important steps in the history of science but also some which constitute unusual and rare sources of scientific thought. This selection has naturally been guided by subjective choice and the richness of the Collection suggested several other approaches, which hopefully will be explored by others.

In my work I benefited from the assistance, advice, and encouragement of many friends and scholars who increasingly shared my enthusiasm for this work and devoted much of their expertise and time to its improvement.

Special thanks are due to Dr Anna Marie Albrektson who, during the spring and summer of 1992, provided me with the most desirable conditions to work with the Collection and helped me to locate most of the material.

I am deeply indebted to Tomas Anfält, who has helped me in the research and transcription concerning Swedish items and has been of constant encouragement.

I would also like to thank Dr Viveca Halldin Norberg and the Staff of the Manuscript Department of Uppsala Univerisity Library who assisted me in many ways during the preparation of the Catalogue.

I am grateful to the Staff of the Photographic Department of the Library who have been extremely helpful and effective in providing this catalogue with its iconographic apparatus.

I thank the Staff of the Maps and Prints Department who provided me with useful information concerning the engravings of the Waller Collection.

While writing the catalogue I received valuable assistance from many scholars. The variety and complexity of the items selected exceeded my competence – had I not been able to exploit the historical, bibliographical, linguistic, and paleographic expertise of the persons who helped me, this catalogue would never have been published. I would particularly like to thank Prof. Ferdinando Abbri, Dr Margarete Andersson-Schmitt, Prof. Ingrid Arthur, Prof. Gunnar Broberg, Dr Antonio Clericuzio, Prof. Allan Ellenius, Ove Hagelin, Prof. A. Rupert Hall, Dr Monica Hedlund, Dr Daniela Majerna, Dr Elio Nenci, Dr Hans Nordesjö, Prof. Nicolaas Rupke, and Dr Sven Widmalm.

I would like to express my gratitude to Prof. Tore Frängsmyr who has supported this project from the very beginning and constantly encouraged me during my work.

Financial assistance at the outset of this catalogue has been generously provided by Marianne & Marcus Wallenbergs Stiftelse and Kabi Pharmacia AB.

Thanks to the enouragment and financial support of Chief Librarian, Dr Thomas Tottie, this book appears in the Acta Series of Uppsala University Library.

The original layout of the Catalogue has been skilfully prepared for the publisher by Xtina Wootz.

Dr Michael Srigley has revised my English and translated the Swedish quotations.

In the realization of this work I owe a great debt to my wife Ilva who provided me with several editorial as well as scientific suggestions which have improved the final result.

Uppsala, February 1993

Marco Beretta

INTRODUCTION

"J'ai un petit cabinet de manuscrits fort précieux & fort chers: quoique je me tue la vue à les lire, j'aime beaucoup mieux m'en servir que des exemplaires imprimés, qui ne sont pas si corrects, & que tout le monde a entre les mains."
Montesquieu, *Lettres Persanes*, in Id., *Oeuvres complètes*, (Paris, 1784), vol. 6, p. 321.

I. MANUSCRIPT SOURCES IN THE HISTORY OF SCIENCE

THE EVOLUTION OF SCIENTIFIC OBJECTIVITY

One of the main characteristics of the natural sciences in comparison with other branches of human knowledge is its explicit aim at openness, objectivity, and universality. By definition science is the state of knowing "as opposed to *intuition, belief,* etc."[4] and its contents can be derived from direct observation or experimentation. In another sense science is "contrasted or coupled with *conscience* emphasizing the distinction to be drawn between theoretical perception of a truth and a moral conviction".[5] Complementary to these general definitions is the common assumption that science is a body of knowledge which can be easily apprehended, provided that one is in possession of its technical nomenclature and of the basic rules of logic.

The universal and objective character of science can be traced back to Aristotle's *Posterior Analytics* in which the definition of it is connected with and founded upon the demonstrative logic orbiting around the syllogistic way of reasoning. Aristotle in fact declared:

> Our contention [...] is that we do at any rate obtain knowledge by demonstration. By demonstration I mean a syllogism which produces scientific knowledge, in other words one which enables us to know by the mere fact that we grasp it.
>
> Now if knowledge is such as we have assumed, demonstrative knowledge must proceed from premisses which are true, primary, immediate, better known than, prior to, and causative of the conclusion. On these conditions only will the first principles be properly applicable to the fact which is to be proved. Syllogism indeed will be possible without these conditions, but not demonstration; for the result will not be knowledge.
>
> The premisses, then, must be true statements; because it is impossible to know that which is contrary to fact, e.g., that the diagonal of a square is commensurable with the sides. They must be primary and indemonstrable, because otherwise we shall not know them unless we have proof of them [...] They must be causative, better known and prior.[6]

The Aristotelian emphasis on the epistemological importance of defining scientific knowledge within the frame of a demonstrative logic had such an

[4] Webster's *New Twentieth Century Dictionary of the English Language,* (New York, 1979), entry "science".

[5] *The Oxford English Dictionary,* (Oxford, 1911), vol. VIII, article "science".

[6] Aristotle, *Posterior Analytics,* I.ii 71b, (London, 1960), p. 31.

impact on the emergence and development of Western science that one might claim that it produced its distinctive character. The appeal to logical demonstration of scientific propositions, the realistic assumption of the existence of the external world and the belief in the possibility of describing it in objective language can be found in most sources of Western scientific literature. However, if it is a fact that the basic rules of Aristotelian logic created the conceptual frame of scientific reasoning, the notions of objectivity, experiment, fact, universality, reality, etc., changed profoundly in their meaning in the course of historical periods. Thus, a plant described in Pliny's *Natural History* according to certain standards of classification and description was perceived differently during the Renaissance when the understanding of a botanical specimen was conditioned by an entirely new approach to reality. This does not imply that the objectivity of science was different for each of the periods considered; rather the criteria of objectivity varied and the two descriptions of the same plant fulfilled the expectations of contemporary readers.

In a famous study on the relation between reality and illusion in art, E.H. Gombrich arrived at the conclusion that representation of reality expressed by artists in various historical periods did not depend on their more or less prominent realistic inclination, but on the general cultural frame of mind within which reality is perceived.[7] In this important work Gombrich gives the following quotation by Riegl in connection with the history of art: "every style aims at a faithful rendering of nature and nothing else, but each has its own conception of Nature".[8] This statement is valid also for the history of science; the Polish physician Ludwik Fleck arrived independently at the same conclusion when he declared: "in science, just as in art and in life, only what is true to culture is true to nature".[9]

The aim of science to be objective and universal was dismissed by Fleck as a positivist illusion and he warned against those historians who tend to measure past science with the logic and epistemological categories of the present. However controversial Fleck's interpretation of the development of the history of science might be, it helps us to introduce a matter which has hitherto attracted little attention, namely that of non-printed sources.

---

[7] E. H.Gombrich, *Art and Illusion. A Study in the Psychology of Pictorial Representation*, (Washington, 1960). Interestingly, in outlining the problem of this work, Gombrich was influenced by the philosophy of science of K.R. Popper.

[8] *Ibid.*, p. 19.

[9] L. Fleck, *Genesis and Development of a Scientific Fact* (1935), (Chicago, 1979), p. 35.

NON-PRINTED SOURCES AND SCIENCE DURING THE MIDDLE AGES

The development of science, like that of any other form of knowledge, has been conditioned by the material conditions under which it developed. Yet, the fact that during the Middle Ages and the early Renaissance scientists used means of communication different from ours has usually been dismissed as an insignificant aspect of their intellectual doctrines.

The standardization of scientific communication and its wide circulation are nowadays familiar in the life of the scientific community but this achievement has been the result of a long and complex historical evolution. Accordingly, when dealing with the scientific thought of the Middle Ages, we have to bear in mind two important elements without which we might fall into grave misconceptions. The first is that when we use the term "science" in relation to the Middle Ages, we use a word and a notion which had no equivalent during that period. The Latin term *scientia* meant in fact the cognitive act of knowledge usually connected with theology rather than the investigation of nature. Secondly, even those branches of natural science which had a recognized identity such as astronomy, medicine, and mathematics, cannot be regarded as scientific disciplines in the modern sense of the word but rather were subsidiary branches of the *Philosophia Prima* or theology which regulated and directed their investigations and their epistemological norms.

The first steps towards a "diffusion" of scientific knowledge were taken in the remote libraries of medieval monasteries, where industrious monks began to transcribe onto thick sheets of parchment the writings of the Greek and Latin authors. During this period, which stretches from around the 6th up to the 12th century, we cannot speak of a diffusion of knowledge, but rather transmission and preservation of the classical sources. It was through the compilation of these *codices* or *libri* that nearly all the surviving Greek and Latin texts in literature, science and philosophy were transmitted. But the cultural function of the monasteries has been of crucial importance for another reason; the monks in fact introduced the techniques of binding the leaves of parchment in a volume *(codex)* which became the first step towards the modern concept of the book.[10]

[10] On this, see the introductory studies of Falconer Madan, *Books in Manuscript. A Short Introduction to their Study and Use*, 2nd ed., (London, 1920); Douglas McMurtrie, *The Book. The Story of Printing and Bookmaking* (1943), (London, 1989), pp. 77-83.

The transcription of the manuscript took place in the monastery's *scriptorium* where the scribes copied the writing and prepared the leaves for the illuminator and rubricator. In spite of the generally accepted rule forbidding any modification of the text copied, the manuscript versions of classical authors transmitted by medieval codices are often full of alterations or deteriorations. Sometimes the copyists not even mastered the language in which the original copy was written, and this easily led to mistakes. Furthermore, when a copy was ready for circulation and contained several mistakes, these were, in their turn, copied by other copyists, and thus the original text gradually deteriorated.

With the emergence of a renewed interest in classical scholarship and humanism during the 13th and 14th centuries, the manuscript versions of the classics transmitted during the Middle Ages were discussed critically and the original text was partially reconstructed. In addition to this, the establishment of universities as organized centres of learning gave a new impetus to the diffusion and renewal of manuscript communication. In this respect, it is worth briefly mentioning the causes of the changing role of the manuscript during the 13th century. Within many European universities the manuscript took on a role and function substantially different from those it had when its production was confined to the monasteries. Scholars, professors, and students demanded "new" material to develop their scholarship and, as a consequence, more copies of the same work were often required. The demand for a faster and more effective production of manuscripts stimulated the creation of the *Pecia,* a system which changed the technique of copying, producing and circulating the manuscript.[11] The latin term *Pecia,* meaning sheep-skin, was applied during the 13th century to the system by which manuscripts were copied and produced in series. A commission established by the university examined the original manuscript that was to be copied and when its text was approved, it authorized the *stationarii* (the booksellers) to lend the manuscript to those who wished to copy it. The original copy, called the *exemplar,* was written on a quaternion, a group of sheets of eight leaves, in the format in-quarto. When there was the demand, the bookseller lent a quaternion of the *exemplar* to the professor, student, or copyist who requested it and, when he had finished with it, he was given a second quaternion by the bookseller and so on until the exemplar had been copied in its cntirety.

---

11 On this, see Jean Destrez, *La Pecia dans les manuscrits universitaires du XIIIe et du XIVe siècle*, (Paris, 1935).

This system of diffusing manuscripts was effective to the extent that in the Statutes of the University of Padua it was stated that the life of the university was dependant on it *("cum absque exemplaribus universitas scholarium stare non possit").*[12] The enormous success of this system significantly changed the attitude of the learned towards manuscripts. Before the introduction of the *Pecia,* manuscripts were rare and their copying was not regulated by any standard pattern. Furthermore, the manuscripts produced in medieval monasteries were not addressed to any specific reader and those manuscripts which did have a circulation were mostly commissioned by the upper classes. The introduction of the *Pecia,* by contrast, anticipated the organized world of the printing press. Its functioning depended on the collaboration of the *stationarii,* the university authorities, the students, the copyists, the rubricators, the illuminators, and the binders. Thanks to this system, a large number of *exemplara* and their copies circulated throughout Europe.

Most of the manuscripts naturally dealt with theology, law, and medicine, that is to say, the main disciplines taught in the university. From our point of view it is interesting to follow the development of one type of scientific manuscript, namely the medical type, and to verify whether the manuscript tradition had any relevant effect on the emergence of a scientific discipline and on the direction of its doctrinal development. An answer to this question, preliminary and tentative as it must be, will be important because it has been a truism among historians of science, as well as among historians of books, to associate the emergence of the scientific revolution of the 15th and 16th centuries with the invention of printing. In an influential work on the development of printing, for instance, it has recently been stated that the cause of the scientific revolution "was not that men turned from books to nature, but rather that books went from script to print".[13] Since the time of Bacon, this association between science and the creation of the printing ateliers has exerted a great influence on our vision of manuscript culture. Justifiable as it might be, this claim had at least one negative effect. It led historians to regard the manuscripts, the *codices,* the rolls, etc., produced during the Middle Ages and the Renaissance as if they were printed books. Accordingly, the fact that before 1450 scientific works were obviously conceived within a system of communication different from the one we have become accustomed to after

---

[12] *Ibid.*, p. 87.

[13] Elizabeth Eisenstein, *The Printing Press as Agent of Change,* (Cambridge, 1979), p. 488.

the introduction of printing has been regarded as irrelevant to the understanding of these same works. The main reason for this attitude is that nowadays we are usually in possession of definitive printed editions of the texts of Antiquity and of the Middle Ages and that the philological and historical investigation of the original handwritten text has become obsolete. Only paleographers, classical philologists, and medievalists are interested in and have the expertise for the study of the manuscript sources of classical and medieval scientific thought whereas most historians of science rely on the printed texts. Thus, the context in which a manuscript was produced, the material used to produce it, the readership it was addressed to, the illustrations and the diagrams it contained are usually separated from its contents as if they were totally independent.

In recent years, this attitude has been partially corrected and some attempts have been made to bring out the characteristics and the role that the culture of manuscripts had in the development of medieval scientific thought.[14]

An element which allows a better understanding of the nature of medieval scientific manuscripts is their iconographic materials.[15] The illustrations that were inserted in these manuscripts emphasized the regulating and directive role of manuscripts in the perception of reality. According to Murdoch, the objects which adorned medieval scientific manuscripts were not observed but "copied from manuscript to manuscript"[16] and they established an authoritative pattern. "More than that, even when the subject may have been observable – as in the case of contemporary scientists, the then-extant instruments used in a science, various flora and fauna, and so on – in the majority of cases the relevant object was not done. Instead, the illustrator constructed his depiction from what was said about the object in a text or, most frequently, from an earlier one found in another manuscript".[17] The opposition between what was known and what was seen was inevitably decided in favor of the former.

---

[14] John E. Murdoch, *Album of Science. Antiquity and the Middle Ages*, (New York, 1984); Linda Ehrsam Voigts, "Scientific and Medical Books", in Jeremy Griffiths and Derek Pearsall (eds.), *Book Production and Publishing in Britain. 1375-1475*, (Cambridge, 1989), pp. 345-402. For medical manuscripts and early printed iconography, see the classical studies by K. Sudhoff, *Tradition und Natur-beobachtung in den Illustrationen medizinischer Handschriften und Fruhdrucke vornehmlich des 15. Jahrhunderts*, (Leipzig, 1907); and Fleck, (1979), *op. cit.*

[15] On this aspect, see Murdoch (1989), *op. cit.*

[16] *Ibid.*, xi.

[17] *Ibid.*, p. 173.

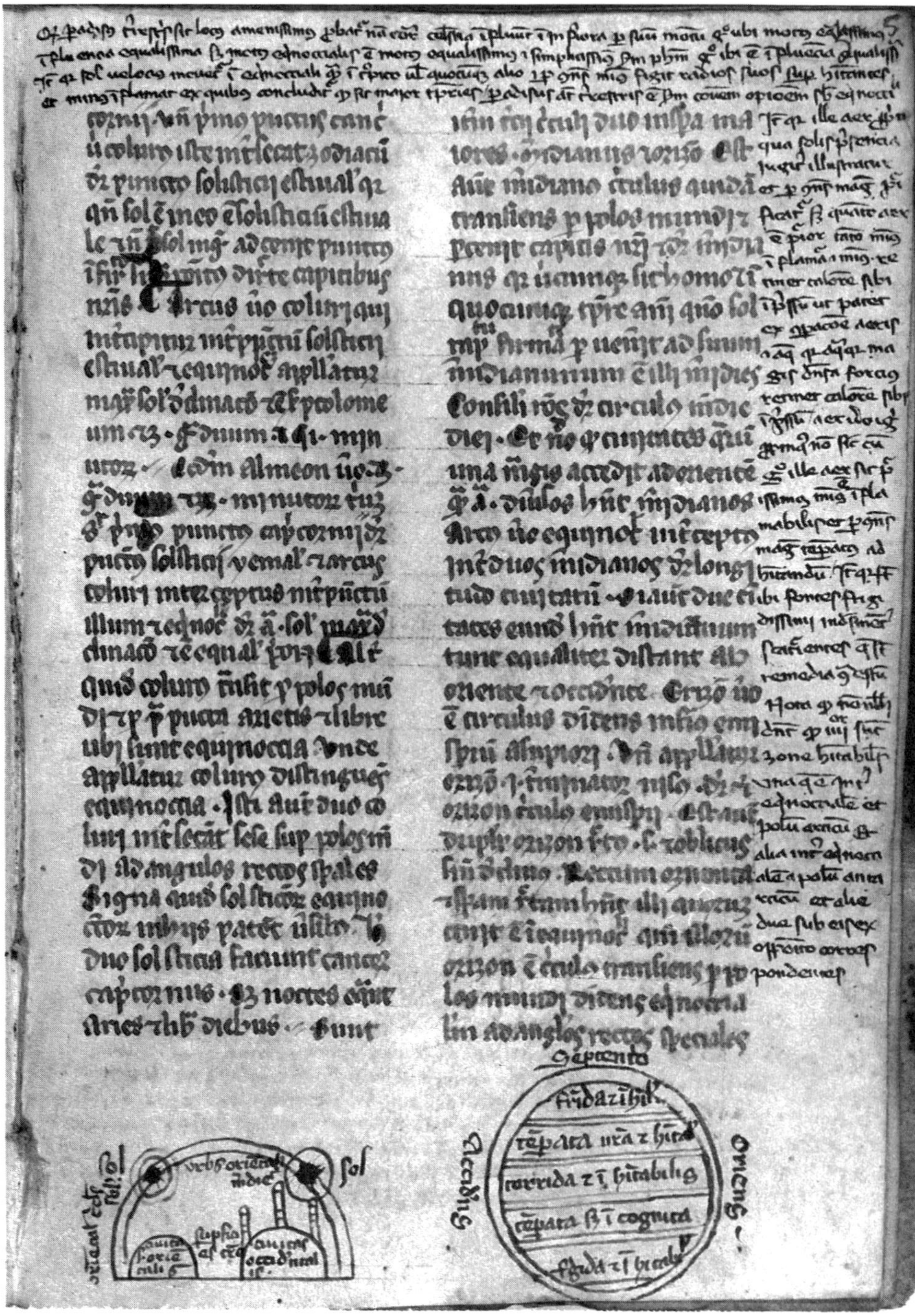

Fig. II. *A leaf of a manuscript's copy of Sacrobosco's* Tractatus de sphaera. *(See Catalogue, item No. 2.)*

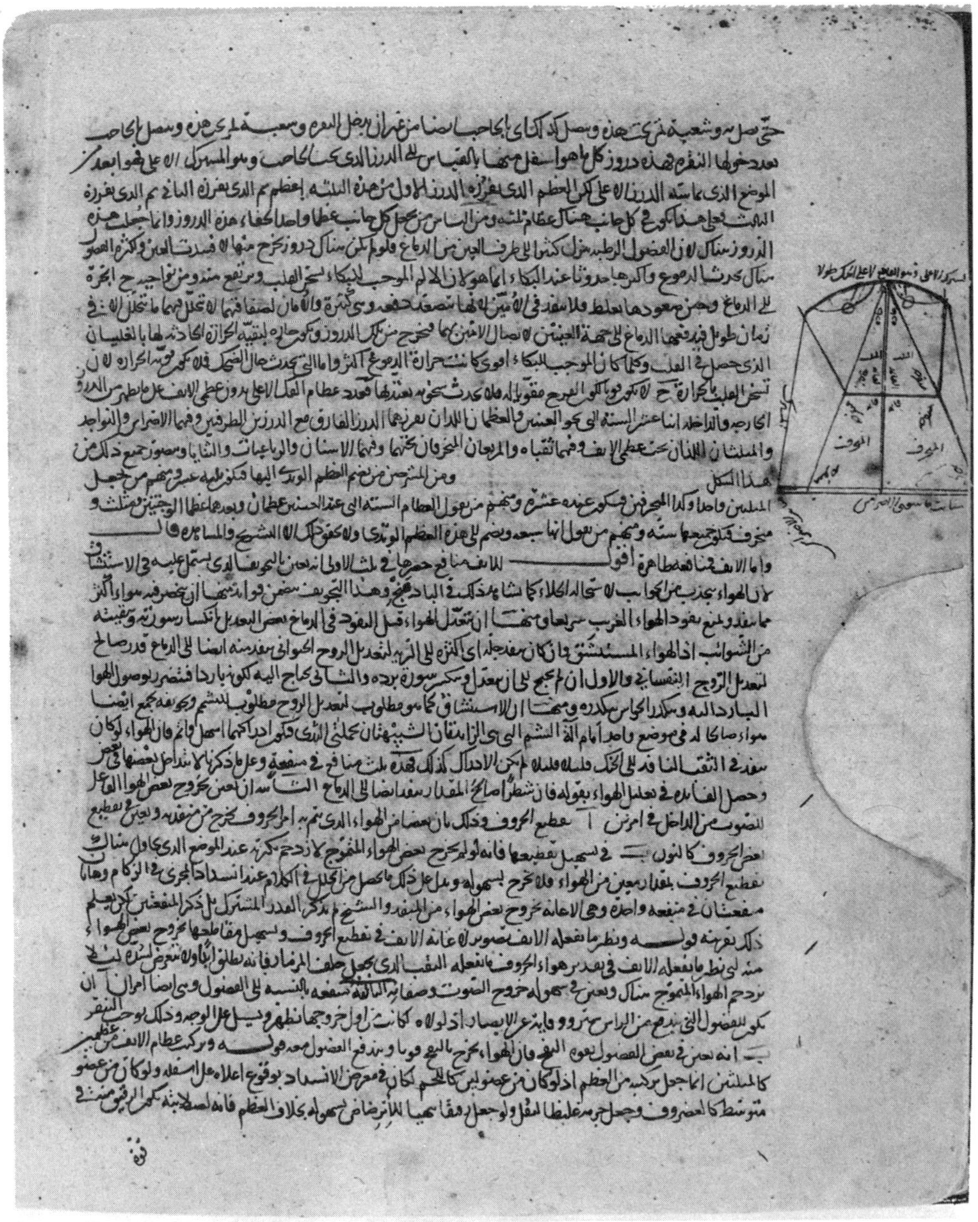

Fig. III. *A leaf of an Arabic commentary on Avicenna's* Canon *dated 1359. (See Catalogue, item No. 3.)*

Thus, the perception of reality was mediated if not affected by the handwritten knowledge of it. This typical trait of the medieval mind shows how little importance was attached to what in modern science is regarded as the sources of any investigation, i.e. observation and experimentation. Murdoch's concise and effective characterization of this medieval trait cannot be bettered: he claims that during the Middle Ages "science was *livresque*. It was not just set down in books; it was largely carried out in books".[18] The handwritten book established an authority which became inconceivable even during the Renaissance. Manuscripts constituted not only the privileged means of scientific communication but they also legitimized and established the authority of medieval culture. The deference of medieval scholars to the written text cannot be separated from the extraordinary prestige and cultural importance enjoyed by the manuscript. The scarcity of manuscripts, their extremely high costs, the beautiful illuminations, etc., give some ground for their authority. In not being separated from theology, medieval medicine and science rested on this kind of authority. "One might say that coming to know about the spread of scientific ideas in Antiquity and the Middle Ages requires that we know about the dissemination of books."[19]

It is hardly surprising then that the learned of the Middle Ages subordinated their empirical knowledge to the authority of the written word, which was regarded as the highest intellectual achievement. Furthermore, the dissemination of manuscripts was still too limited to encourage a critical attitude towards the scientific relevance of their contents. Medieval medicine, astronomy, mathematics, physics, alchemy, etc., still depended to a great extent on the hegemonic handwritten canonization of the classical tradition and its passive transmission. Natural phenomena were not comprehensible unless there was an existing written explanation of them within which they might be conceived and therefore perceived. Accordingly, scientific problems which historians have found "modern", such as those related to the plurality of worlds and infinity, were not leading medieval philosophers to the kind of questions which were raised during the 16th and 17th centuries.

This "limitation" was not due to the lack of an intellectual background or of technical notions, but to the fact that within their philosophy the question of the infinity of the universe belonged to a different epistemol-

---

18 *Ibid.*, p. 3.

19 *Ibid.*

ogical tradition.[20] In the case of medicine, for instance, a medieval physician represented the human body according to the established written tradition, and from his point of view this was the pattern through which he approached the *real* picture of the human body. A faithful representation of reality depended more on the description one could find in a written text than on the observable features of it.

The importance of manuscripts had such an impact on the development of science that early printed books imitated, typographically and iconographically, the features of the manuscripts,[21] and during the 16th century it often happened that printed books were given to the rubricator and illuminator as if they were manuscripts.

THE RENAISSANCE OF LETTERS (1450-1800)

With the introduction of printing and the systematic establishment of printing ateliers throughout Europe, the role of the manuscript in science, as well as in other branches of knowledge, rapidly changed but, contrary to what may be expected, it did not lose its importance. The manufacturing of books was in its infancy an extremely slow and expensive enterprise and only few printing firms were able to produce more than 300 copies of one text before 1470.[22] Another important factor which did not deprive the manuscript of its important function was censorship. Printing firms in fact needed a licence from the university, the Church or local authorities to publish a certain text. As a consequence of this, control over manuscripts rapidly diminished and they circulated freely without any form of official regulation.

However, what changed the nature and the contents of scientific manuscripts was the emergence of a renewed philosophy of nature. It is obviously beyond the task of this essay to approach such a complex theme as the phi-

20 In his "Le vide et l'espace infini au XIVe siècle"(1949), Alexandre Koyré had persuasively argued that the discussion of infinity and the plurality of worlds of the 13th century was not, as Pierre Duhem claimed, an anticipation and a source of the modern cosmology which followed Copernicus' work, but an integral part of the theological debates and it was subordinated to the authority of scholastic philosophy. See Koyré, *Études d'histoire de la pensée philosophique*, (Paris, 1971), p. 37ff.

21 On this, see N.F. Blake, "Manuscript to Print", in Jeremy Griffiths and Derek Pearsall (eds.), *op. cit.*, pp. 403-432.

22 C. Schmitt et al. (eds.), *The Cambridge History of Renaissance Philosophy*, (Cambridge, 1990), p. 28ff.

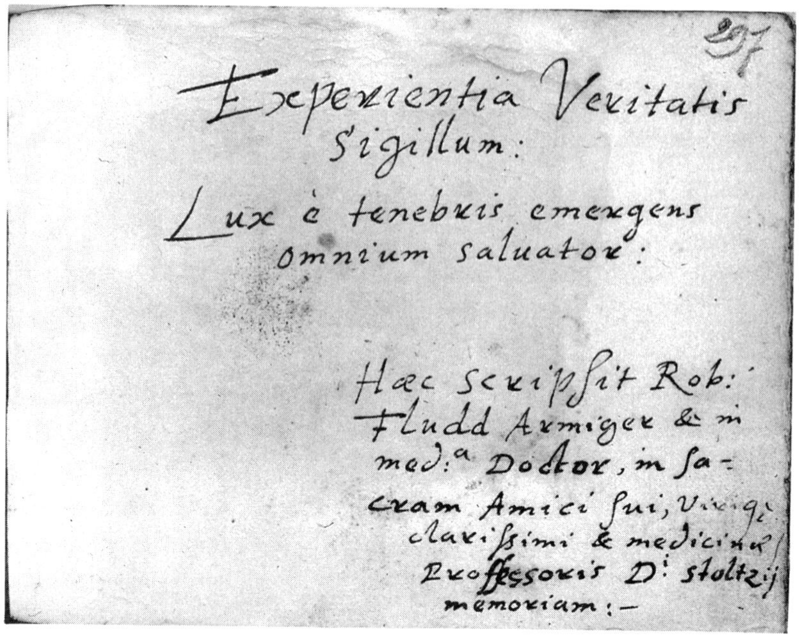

Fig. IV. *A leaf of the Album amicorum belonging to Daniel Stolz von Stolzenberg bearing a dictum and the signature of the English alchemist Robert Fludd. (Uppsala University Library, shelfmark: Y 132d.)*

losophy of nature during the Renaissance[23], but it is worth mentioning the aspects of this renewal which are important for the understanding of the change of manuscript culture during the 15th and 16th centuries.

Many natural philosophers of the Renaissance were inspired by an open hostility to the Aristotelian philosophy of science, to its logic, its emphasis on the authority of the senses, etc. The Aristotelian representation of nature which offered an ordered picture of the universe and, above all, which imposed strict limits on the power of the human intellect, was rejected during the Renaissance and it was replaced by an intellectual confidence which allowed men to think that everything was possible and everything understandable. The fact that the writings of Aristotle were beginning to exert their authority through printed books provided the manuscript, as we shall see later, with a new role.

---

[23] Among other studies, see P.O. Kristeller, *Renaissance Thought and its Sources,* (New York, 1979); R. Lenoble, *Esquisse d'une histoire de l'idée de nature,* (Paris, 1969), pp. 279-307; A. Koyré, *From the Closed World to the Infinite Universe,* (Baltimore, 1957); George Sarton, *Six Wings. Men of Science in the Renaissance,* (London, 1957); E. Garin, *Scienza e vita civile nel Rinascimento,* (Bari, 1965).

The philosophers of the Renaissance were able to find and explore new paths and established new cognitive connections between the Microcosmos and the Macrocosmos. Accordingly, the demonstrative logic of Aristotle was replaced by alternative forms of reasoning which originated in Hermetic philosophy, alchemy, astrology, and other mystical beliefs. As a consequence, natural philosophers of the Renaissance rejected what was regarded, and officially recognized, as science and constructed methods and doctrines in opposition to it. As a a great historian of science expressively puts it:

> C'est l'époque de la superstition la plus grossière et la plus profonde, une époque où la croyance à la magie et à la sorcellerie s'est étendue d'une manière prodigieuse, et a été infiniment plus repandue qu'au Moyen Age; et vous savez bien que l'Astrologie joue à cette époque un rôle beaucoup plus grande que l'astronomie.[24]

The personality and the doctrines of Paracelsus offer perhaps one of the most representative examples of the philosophy of nature of the Renaissance.[25] By examining critically the classical sources of medical literature, Paracelsus aimed at creating not only a new approach to healing but also an entirely new cosmology. The ontological commitment behind Paracelsus' philosophy of medicine centered on his belief in a correspondence between the Microcosmos and the Macrocosmos and the analogies which structured their mutual relations. This belief projected a sort of omnipotence on the cognitive powers of the human mind. In certain respects man was even superior to nature,[26] always capable of going beyond the limitations imposed by the materialistic world. This credo, which combined natural knowledge with mystical experience and magical beliefs, led Paracelsus to despise the book and to regard it as the symbol of the culture he aimed to overturn:

---

24 A. Koyré, "L'apport scientifique de la Renaissance"(1949), in Id., *Études d'histoire de la pensée scientifique,* (Paris, 1973), pp. 50-51.

25 On Paracelsus, see the two classical studies by A. Koyré, *Mystiques, spirituels, alchimistes du XVIe siècle allemand,* (Paris, 1971), pp. 75-129; Walter Pagel, *Paracelsus. An Introduction to Philosophical Medicine in the Era of the Renaissance,* (Basel, 1982).

26 "Nature emits light, and by its radiance she can be known. But in man there is still another light apart from that which is innate in nature. It is the light through which man experiences, learns, and fathoms the supernatural. Those who seek in the light of nature speak from knowledge of nature; but those who seek in the light of man speak from the knowledge of super-nature. For man is more than nature; he is nature, but he is also spirit, he is also an angel, and he has the properties of all three", from Paracelsus, *Selected Writings,* (Princeton, 1988), pp. 43-44.

> The book in which the letters of the mysteries are written visibly, discernibly, tangibly, and legibly, so that everything one desires to know can best be found in this self-same book, inscribed by the finger of God; the book compared with which, if it is correctly read, all other books are nothing but dead letters-know that this book is the book of man, and should not be sought anywhere but in man alone. Man is the book in which all the mysteries are recorded; but this book is interpreted by God [...]
>
> For it is not on paper that you will find the power to understand, but in Him who put the words on paper.[27]

Truth can be found only by exploring the soul of man and the knowledge of the natural world primarily relied on this innate capacity. Not surprisingly, this mystical belief led Paracelsus to an explicit criticism of the livresque tradition of knowledge and more specifically of the Aristotelian philosophy of nature. The printed books, with their connections with the Universities, their system of privileges, licences, etc., was for him the symbol of the cultural establishment. Furthermore, in Paracelsus' eyes the dead authority of the written word was opposed to the vitality and vivid impressions of the active observation of nature. Finally, the belief of traditional scientists that truth was an *adequatio intellectus rei* was drastically overturned and the complexity and variety of nature was subordinated to the spiritual and intellectual power of a few individuals.

As for the method, Paracelsus introduced another break with the Aristotelian tradition:

> We men discover everything that lies hidden in the mountains by external signs and correspondeces, and thus also do we find all the properties of herbs and everything that is in the stones. There is nothing in the depths of the seas, nothing in the heights of the firmament, that man is unable to discover. No mountain, no cliff, is so vast as to hide or conceal what is in it from the eyes of man; it is revealed to him by corresponding signs.[28]

The cognitive omnipotence of man was a quality which was neither common to all (as the common sense of Descartes), nor something that could be comprehended. It was an extremely rare gift and difficult to cultivate. In order to do so, the scientists needed a thorough knowledge of chiromancy, magic, physiognomics and, above all, alchemy. These occult sciences provided him with the means to go beyond the fallacy contained in the empirical knowledge of nature. This method of investigation was in total opposition to that of Aristotle who based his philosophy on the observable and empirical

---

27 *Ibid.*, p. 44.

28 *Ibid.*, pp. 120-121.

qualities of natural objects; Paracelsus overturned the privileged role attributed by Aristotle to the ratiocinative part of the human intellect. To him "magic has power to experience and fathom things which are inaccessible to human reason. For magic is a great secret wisdom, just as reason is a great public folly".[29]

The opposition asserted by Paracelsus between secret and public knowledge could not but have significant effects on the role of the manuscript during the Renaissance. In effect, a flood of manuscripts dealing with occult sciences were in circulation during the 15th and 16th centuries. The belief that the natural knowledge of things was accessible only to a relatively small number of adepts led to the use of manuscripts as the privileged means of communication. It is certainly not by chance that naturalists such as Kepler, Newton, and Leibniz, who dealt with alchemy, decided not to publish their manuscripts. This was a deliberate choice that confirmed the occult and secret nature of alchemy.

If, on the one hand, there is little doubt that manuscripts assumed a prominent function among those naturalists who cultivated occult sciences, it would not be correct, on the other hand, to confine their importance only within the limits of that tradition. As a matter of fact scientific manuscripts fulfilled other requirements for the naturalists of the Renaissance. Among these two can be mentioned: 1) "the exact reproduction of complicated designs of the heavens required a technological advance which was not available in the first century of printing. Printed designs and drawings could not reproduce details as clearly and as accurately as needed, so scholars had to rely on hand-made copies in order to secure the fidelity and accuracy demanded by more complicated scientific study".[30] Indeed the emphasis put by Renaissance naturalists on the importance of direct observation of nature required a faithful representation of it. The art of observation was emancipating itself from the abstract forms of representation of the medieval manuscripts and when it came to putting it on paper, new techniques, such as the invention of perspective, were needed. 2) Another important aspect of manuscripts during the Renaissance was their rapid and free circulation. The exchange of letters between naturalists became an important means of communication during the 16th century and it created among them an emerging self-confidence as a scientific community. This form of correspondence intro-

---

29 *Ibid.*, p. 137.
30 J. D'Amico in C. Schmitt et al. (eds.), *op. cit.*, p. 24.

duced some new elements which had been unknown to medieval scholars. The renewed interest in observation, the enormous curiosity which followed it, and the geographical discoveries of the 15th century, led naturalists to exchange not only letters but also natural specimens, books, manuscripts, drawings, engravings, etc., in order to satisfy their thirst for knowledge.[31] This exchange became so common among naturalists that several specimens described in works on natural history had never been observed by the authors themselves but knowledge of them was based on the information received from scholars who had actually seen the specimens. Such was the case with several birds described in Aldrovandi's *Ornithologia* (1603), or, later, with many plants described in Linnaeus' *Systema naturae.* A naturalist observed and studied the flora and fauna of an unexplored area and when he returned shared his knowledge with his colleagues by sending them information and some of the material he had collected. In return he naturally expected to be acknowledged and to get the same kind of treatment when the occasion would come. This kind of exchange of scientific information, carried out mostly through correspondence, remained important until the end of the 18th century, that is to say, until scientific journals became the main source of information.

During the 17th century scientists developed a self-confidence in their intellectual powers and in the potentialities of their disciplines which was unknown to their predecessors. New societies and academies were founded throughout Europe to promote the cultivation and development of the sciences. University professors in mathematics and medicine became more and more independent from the control of the Faculty of Theology, a fact which encouraged new philosophical investigations. Naturally, this process was neither as linear nor as rapid as is usually depicted in the manuals of the history of science. The professional scientist who could afford to live off his work was still a rare exception – prominent figures such as Descartes, Spinoza, Leibniz, Galileo, Kepler, a.o., depended on the generosity of the patronage of kings, courts, and friends. However, their awareness of the goal in view and of the cultural importance of science in society was surprisingly advanced. Scientific knowledge was perceived as a distinct part of human knowledge with a distinctive method of reasoning. In a few cases, as for instance in Descartes, this awareness was transformed into an entirely new

[31] An interesting study on this topic is that by Giuseppe Olmi, "<<Molti amici in varij luoghi>>: studio della natura e rapporti epistolari nel secolo XVI", in *Nuncius,* VI:1 (1991), pp. 3-31.

philosophy in which its method and contents were based on the achievement of natural science in general and mathematics in particular.

As a consequence of the emergence of modern science, the role of the manuscript began to change accordingly. At this stage the distinction between the role played by manuscripts and that played by correspondence became pronounced.

During the 17th century the scientific manuscript had different functions, the most significant of which was probably its use to byepass the censorship of the Church. A famous example of this kind is given by Descartes. On June 23, 1633, Galileo was condemned by the Catholic Inquisition for his scientific defense of the Copernican system of the universe. From 1629 onwards Descartes had been working on a treatise on physics (*Monde)* in which he was also defending the Copernican system. Fearing that he might be dragged into the same critical position as Galileo, Descartes decided not to publish his manuscript which remained incomplete and was only published posthumously some thirty years later. Its late publication deprived Descartes' work of the impact he hoped for and most of its scientific contents had already been superseded by new contributions to the subject.

Another example shows a similar situation but a different fate. Alessandro Marchetti, Professor of Mathematics at Pisa University, pupil of the physicist, Giovanni Alfonso Borelli, and prominent follower of Galileo, decided in the 1660's to translate Lucretius' *De rerum natura* into Italian. The poem, as is known, combined scientific knowledge with an atomistic and materialistic view of the universe. Furthermore, Lucretius' attack against religion made *De rerum natura* the most appealing work to use as a defence of the values of the renewed scientific spirit. Marchetti finished his translation in 1669 and in 1670 he asked the ecclesiastical authorities of Florence for permission to publish it.[32] Ultimately, the licence was denied and the Grand Duke of Tuscany, Cosimo III, "returned the manuscript to Magliabechi [the Court librarian] with an expression of surprise that the author should have employed his spare time so ill".[33]

The negative response of Cosimo III and of the Church authorities immediately raised curiosity among those naturalists, philosophers, and freethinkers who were engaged in the defense of an "enlightened" society. As a

---

32 On this, see Cosmo Alexander Gordon, *A Bibliography of Lucretius,* (London, 1962), pp. 194-200.

33 *Ibid.*, p. 195.

result Marchetti's translation, which included a few verses in defense of Gassendi, had a great success and several manuscript copies were made. Its success is evidenced by the fact that several prominent thinkers of the early Enlightenment such as Bayle, Leibniz, and Voltaire possessed a copy of it. In 1717 Marchetti's translation, which was already at the height of its success, was finally published in London and in 1718 it was placed on the Index together with all manuscript versions. As a consequence of this persecution, Lucretius became one of the most popular poets among the *Philosophes*. D'Holbach and Diderot promoted a famous French translation of it and several scientists opened their works with a verse taken from *De rerum natura*. The fate of Marchetti's work was not at all uncommon between the 16th and 18th centuries as many philosophers and scientists preferred to keep their controversial papers in manuscript form. Furthermore, the huge production and the wide circulation of clandestine literature during that period shows that the manuscripts far from hindering the diffusion of a thought did in many cases actually attract the attention of the learned world.

Another form of scientific manuscript which became increasingly common during the 18th century was that of the memoirs or papers read before or deposited in scientific academies which, though containing original discoveries, were not yet ready for publication. Lavoisier offers a famous example of this kind of scientific manuscript. On November 2, 1772, Lavoisier deposited with De Fouchy, the Secretary of the *Académie Royale des sciences* of Paris, a sealed note containing a concise description of his experiments on the combustion of phosphorus and sulphur.[34] Some of the contents of this short manuscript are worth quoting:

> About eight days ago I discovered that sulphur, in burning, far from losing weight, on the contrary gains in weight [...]
>
> This discovery, which I have established by experiments that I look upon as decisive, has led me to think that what is observed in the combustion of sulphur and phosophorus may well take place with regard to all substances that gain in weight by combustion and calcination: and I am persuaded that the increase in weight of metallic calces is due to the same cause [...]
>
> This discovery appearing to me one of the most interesting of those that have been made since the time of Stahl, I felt compelled to secure my right

[34] The historical significance of this note has been examined by A.N. Meldrum, *The Eighteenth Century Revolution in Science. The First Phase,* (Calcutta, 1930), p. 3ff. and by Henry Guerlac, *Lavoisier – The Crucial Year: The Background and Origin of His First Experiment on Combustion in 1772,* (Ithaca-New York, 1961).

> in it, by depositing this note in the hands of the Secretary of the Académie, to remain sealed until the time when I shall make my experiments known.[35]

Historians of chemistry agree that this text is of extreme importance for a correct understanding of the development of Lavoisier's revolutionary chemical theory of the gaseous state, which ultimately changed the face of 18th century chemistry. However, they have faced some serious difficulties which have not been solved satisfactorily. Too little attention has been paid to the form and the circumstances which led Lavoisier to leave his note with De Fouchy. But let us examine the paper in detail.

The effects of the experiments on phosphorus and sulphur were already known and many predecessors of Lavoisier had already observed and published the same phenomena. The observation that metals gained weight when calcined was traceable back to Biringuccio's *De la pirotechnia* which was published in 1540! What was then the discovery that Lavoisier was so eager to ensure for himself? The answer is not so simple. From a literal reading of the text it was the idea that the combustion and the calcination of chemical substances led to their gain of weight and that the cause of this gain had to be the same, i.e. the combination or the fixation of the air in these substances. A contemporary reader of the contents of this text would have found it either not very original or too vague for the magnitude of the discovery to be grasped. In addition, some historians who have read the text in the same way have concluded that Lavoisier did not discover anything particularly new. However, if this was the case, why did a careful and accurate scientist like Lavoisier take the step of ensuring for himself, with such self-confidence, priority in an important discovery when in fact it was already known? In reality, the manuscript contains an intuition rather than an account of an experiment. Lavoisier was beginning to think that all substances were subject to calcination and combustion combined with a gas and that therefore the explanation of the most common operations of chemistry could be reduced to a simple cause. As early as 1772, Lavoisier was still far from the isolation and the definition of gases and his understanding of their chemical role in a specific reaction was not as sharp as that showed by English naturalists in their works. As a matter of fact, Hales, Black, and Priestley were more advanced at the experimental level than Lavoisier. However, instead of conceiving science as an accumulation of facts, the French chemist followed the Cartesian method and began his career in chemistry by setting forth an hypothesis

35 Quoted and translated by Meldrum, *op. cit.*, p. 3.

Fig. V. *A leaf of the Album amicorum belonging to the German physician Johann Gastel on which a caricature of 16th century physicians is drawn.*
(Uppsala University Library, shelfmark: Y 55.)

which did not have any firm experimental ground. As Meldrum correctly puts it, the project Lavoisier set out in the sealed note of 1772 became a "fixed idea"[36] around which he built up a new system of chemical theory. Therefore, while he was ensuring himself the right of the discovery by means of the sealed note, Lavoisier was aware of the theoretical consequences his research project would eventually lead to but he did not yet have the experimental basis to prove it. Aware that science proceeded by the experimental testing of a theory and by its predictive power, Lavoisier wanted to wait and see whether his intuition would be right. Furthermore, good tactician as he was, he understood that if he had submitted his discovery to the public as it stood in 1772 he would not have had the response it deserved. Thus, the fact that Lavoisier did not publish his paper does not diminish its importance. On the contrary, the eagerness shown by Lavoisier to get full credit from the Academy of Science for his discovery, stresses the importance of this source. The sealed note became in fact the unpublished nucleus around which he built his theory

[36] *Ibid.*, p. 6.

of oxygen. This is confirmed by the fact that the sealed note was the only manuscript by the French chemist which was submitted to such a formal procedure. All his discoveries and experiments such as the decomposition of water, the investigation of heat, and the experiment on the respiration of animals, were recorded in the laboratory notebooks, and the vindication of their priority was made by Lavoisier only through ordinary publication.

However relevant the role of the manuscript might have been in single cases during the 17th and 18th centuries, its role and importance in science generally declined. The most characteristic form of handwritten scientific work during this period remained the exchange of letters. The letter in fact not only constituted for scientists the fastest means of communication and collaboration, but it also represented a form of assessing in a public way, even if not yet officially, the priority of a discovery or of an experiment. Particularly during the 17th century, when the publication of a book was an expensive and often slow enterprise, scientists communicated their experimental findings through letters. Letter-writing became a public way of communication and, following a very common routine, relevant scientific letters were copied and made to circulate throughout Europe. As shown in the example described in item No. 19 of this Catalogue, Galileo transcribed a letter by Kepler and had it circulated. Likewise his pupil Evangelista Torricelli (item No. 27) explicitly asked his addressee to copy the letter in which he made an original geometrical demonstration and to make it known to other scientists in order to assess the priority of his discovery.

The importance of scientific correspondence is shown by the enormous growth of this means of communication. The voluminous correspondence maintained by Mersenne, Oldenburg, Malpighi, Descartes, Galileo, Kepler, Newton, a.o. shows how common and, at the same time, how important the exchange of scientific letters was. Indeed, it is impossible to have a correct and complete understanding of 17th century science without taking the medium of correspondence into consideration. This is less true with respect to the 18th century, in spite of the fact that scientists wrote far more letters than their predecessors.[37] During the Age of Reason writing letters became a successful

37 It is enough to mention the monumental correspondence of Linnaeus, Haller, Spallanzani, and Euler. On 18th century scientific correspondence, see Ferdinando Abbri, "<<Commercio di libri e produzioni naturali>>. Lettere di Giovanni Mariti e Attilio Zuccagni a Carl Peter Thunberg", in *Nuncius*, VII:2 (1992), pp. 119-139; Sven Widmalm, "A Commerce of letters: Astronomical Communication in the 18th Century", in *Science Studies*, 2 (1992), pp. 43-58.

literary form. Thanks to the works of Voltaire, Montesquieu, and Rousseau, a great number of philosophical treatises as well as literary works were composed in the form of fictional correspondence. This epistolary form also affected scientific literature and it is enough to mention the works by Euler, Diderot, Bailly, Cassini, Magalotti, and Volta to show how successful the use of this literary form became among scientists.

The literary success of correspondence combined with decreased costs of printing led scientists to write letters to real addressees with a view to their possible publication as a separate work.[38] The public nature of scientific letters was so taken for granted that they were often published without the permission of the sender. In addition to this, the diffusion and the success of scientific journals, periodicals, and academic transactions created an ideal space for the publication of scientific letters. It is enough to dip into these journals to grasp the proportion and success of this form of publication. More often the letters, which contained some original information on a scientific topic, were read before the scientific academies or societies the receivers belonged to and eventually published in the academic transactions. As a consequence of this organized system of communication an original scientific discovery rarely failed to reach public attention.

Another form of scientific manuscript which is relevant to the historical understanding of the thought of scientists are the laboratory notebooks. Usually, these records contained the experimental data collected in a laboratory, the observations collected in the field etc., providing a rough idea of the sources utilized by a scientist during his investigations.[39] This material was often subordinated by the scientists to the consistent reasoning of their synthetic works which were eventually published. They are, however, important to historians because they illustrate the increasing difficulties scientists had to face in elaborating at a theoretical level the empirical material they had collected. In the case of natural history, the notebooks or the journals were often accompanied by drawings depicting the flora and the fauna, which constitute an important but often neglected source concerning the development of the method of clarifying and perceiving the natural world.[40]

---

[38] There is, for instance, the case of the Swiss geologist, Jean André De Luc, who published several of his works after his correspondence.

[39] On this, see David Knight, *Sources for the History of Science. 1660-1914,* (London, 1975), pp. 87-90.

[40] See the papers collected in the forthcoming work edited by R. Mazzolini, *Non-Verbal Communication in Science Prior to 1900,* (Florence, 1993).

During the 19th century manuscript sources still represented an important aspect of the everyday life of a scientist, but the increasing standardization of scientific communication and the powerful position achieved by journals subordinated them to printed papers and books. Manuscripts became the preliminary source of a printed work and scientists became extremely aware of the distinction between the role of a manuscript, a laboratory record, a journal, and correspondence on the one hand and the printed works on the other. The book and the article increasingly became the only form of public communication among scientists. This is true in even greater measure in the 20th century when the culture of the manuscript has practically disappeared. Today, it is, to say the least, unusual that scientific ideas circulate in the same way they did during the 16th and 17th centuries. The international scientific community has subordinated the use of manuscripts to the printed word which means that a paper has no scientific value until it is published. The economic implications of assessing the priority of a scientific discovery have led scientists to speed up enormously the publication of their papers. As recently shown by the claimed discovery of nuclear cold fusion, scientific ideas become public and are published even before they have been submitted to the most elementary experimental tests. As a consequence of this rush into print, the manuscript has not only lost any technical reason for existing but, more importantly, it has lost the intellectual background which has supported its existence for nearly two thousand years.

II. The Waller Collection

Erik Waller (1875-1955)

Erik Waller was born in Rangeltorp near Önum in 1875.[41] After finishing his schooling in 1894, Waller began to study medicine at Uppsala University and obtained his medical degree in 1900. Soon after he specialized in surgery at the "Karolinska Institutet" in Stockholm. His career as a surgeon was brilliant but not exceptional. In 1909 he became surgeon-in-chief at the hospital in Lidköping where he continued to practise until his retirement in 1933. During the period 1933-1940 he probably practised as a private physician but there is evidence that he began to be interested in deepening his knowledge of the history of medicine by attending international conferences[42] and that he visited several European cities. As a result of his increasing historical interests, Waller presented in 1935 at the XIth International Conference of the History of Medicine held in Madrid a paper on Vesalius' Epitome of the *De Humani corporis fabrica*. This paper, which was eventually published in *Lychnos*,[43] described a unique edition (Basel, 1555) of this work which was in Waller's possession (Fig. VI). Unfortunately this was the only historical work published by Waller. In spite of the scarcity of printed documentation, it emerges quite clearly from a close examination of his manuscripts and letters that Waller was well acquainted with the history of medicine. Many prominent historians of medicine of the first half of this century such as Karl Sudhoff,

---

41 The information concerning Waller's biography is not particularly rich. However, some valuable reconstructions of his life are those by H. Sallander, *Bibliotheca Walleriana*, vol. 1, (Uppsala, 1955), xi-xii; John Fulton, "Axel Erik Waller 1875-1955", *Journal of the History of Medicine and Allied Sciences*, 10 (1955), pp. 226-227; T. Kleberg, "Erik Waller", *Libri*, 6 (1955), pp. 76-79; Sten Lindroth, "Erik Waller", *Bulletin of the History of Medicine*, 30 (1956), pp. 88-90; H. Krook, "Samlare och Konässör – Om Erik Waller och hans medicinhistoriska bibliotek", *Nordisk Medicinhistorisk Årsbok*, (1976), pp. 1-8.

42 Waller attended the Xth and XIth International Congress of the History of Medicine where he met several historians of medicine who eventually began to ask him for bibliographical information or transcriptions of letters.

43 In 1937 Waller published "Eine unbekannte Ausgabe von Vesals Epitome", *Lychnos*, (1937), pp. 251-260. The work by Vesalius is *De humani corporis fabrica librorum epitome*, (Basel, 1555). This unique copy belonged to the Italian 18th century anatomist, Antonio Scarpa, and it was bought by Waller from an Italian bookdealer. Enclosed in the book are several documents concerning the rarity and the history of this edition, (see Sallander's *Bibliotheca Walleriana*, no. 9909).

Fig. VI. *The frontispiece of the unique extant copy of Vesalius'* De Humani corporis fabrica librorum epitome, *(Basel, 1555).* (Bibliotheca Walleriana, *no. 9909.*)

George Keynes, Henry Ernst Sigerist, Arturo Castiglioni, John Fulton and many others, exchanged letters with Waller and often asked him to help them with his bibliographical expertise. Waller also exchanged letters with the most important historians of science of his time, i.e. George Sarton, Charles Singer, and Aldo Mieli[44] and he was internationally recognized not only as the most important collector of the time but also as an expert historian of medical literature.

It was probably due to his bibliographical competence that in 1940 the Swedish Medical Society in Stockholm offered him a post as librarian. As the library of the Society stored a rich collection of rare medical books,[45] Waller accepted the position which he retained until 1946.[46] In 1946 he returned to Lidköping where he continued to collect books and manuscripts dealing with the history of science and medicine until his death on January 28, 1955.

THE BOOK COLLECTION: A REASSESSMENT

Thanks to the untiring effort of Chief librarian Tönnes Kleberg, who promised him that he would publish a catalogue of his book collection, Waller decided to donate his library to Uppsala University Library. The publication of the catalogue took several years and it was only in 1955, a few months after Waller's death, that Hans Sallander published his *Bibliotheca Walleriana*.[47] After the publication of the Catalogue, the Waller Collection became immediately known all over the world and the Catalogue soon established itself as the most important bibliography of medical history. The Collection is indeed extraordinary and it is probably the largest of its kind dealing with the history of science and medicine ever collected by an individual. The Catalogue consists

44 The correspondence of Waller offers an extremely interesting source for historians of medicine as it contains many bibliographical items of information concerning medical books and manuscripts. The correspondence is kept in one box with the shelfmark Waller 653 B:1.

45 Two select catalogues of the holdings of the library are given by Ove Hagelin in his *Rare and Important Books in the Library of the Swedish Society of Medicine*, (Stockholm, 1989) and *The Byrth of Mankynde otherwyse named The Womans Booke*, (Stockholm, 1990).

46 A claim based on oral tradition and which has enjoyed general credit suggests that Waller took from the library some rare items missing in his own collection. Apart, however, from a few but not rare titles which he took with the permission of the Society, such a claim seems to be totally unfounded.

47 Sallander, *Bibliotheca Walleriana*, 2 vols., *op. cit.*

of more than 21,000 titles but it does not include some 4000 small prints published in the 19th and 20th centuries.[48] In the Catalogue 150 incunabula are listed, 24 of which were printed before 1480. Among them, there are some that are extremely rare and there is a unique copy of *La vie de Saint Roch* (c.1496).[49] The collection lists many other rare items printed during the 16th century, such as for instance almost unique copies of works by Vesalius, Paracelsus, Saliceto, Lanfrancus, and Fracastoro. Waller also had the largest private collection of editions of works by Ambroise Paré (45 titles), William Harvey (43 titles), Paracelsus (44 titles). With respect to the medical literature printed during the 17th and 18th centuries, the richness of the Waller Collection defies even a rough description.

The collection, however, not only consists of medical books. Most of the works which had a relevant impact on the development of western science are in fact listed. Accordingly, we find the first editions of Copernicus' *De revolutionibus orbium coelestium*, Kepler's *Epitome astronomiae Copernicanae*, Galileo's *Dialogo sopra i due massimi sistemi del mondo*, and Newton's *Principia mathematica*. No less significant is the collection of chemical and alchemical literature which contains many items not included in Duveen's and Ferguson's catalogues.[50] Waller also possessed rare editions of important works in the history of physics, botany, zoology, geology, mathematics and the occult sciences.

The Waller Collection is also unique in another respect. It lists in fact a large number of books with ex-libris, inscriptions and marginalia of important scientists. Among these there are books coming from the libraries of Aldrovandi, Banks, T. and C. Bartholin, Bernard, Berzelius, Bichat, Blumenbach, Boerhaave, Bonnet, Borelli, Buffon, Camper, Charcot, Cullen, Dee, Della Porta (Fig. VII), Fabricius, Fracastoro, Gesner (Fig.VIII), Harvey (see Cata-

---

48 There are 88 boxes of offprints mostly dealing with medicine which have not been catalogued. The reasons for this exclusion from Sallander's Catalogue are not known.

49 Valuable information on the Waller Collection of books may be found in the following studies: Emil Starkenstein, "Der Arzt und sein Buch", *Philobiblon*, 10 (1938), pp. 305-334; Sten Lindroth, "Erik Wallers Bibliotek", *Svenska Bibliotek*, (1946), pp. 3-58; Hans Sallander, *The Bibliotheca Walleriana in Uppsala University Library*, (Uppsala, 1951); Hans Sallander and Tord Skoog, *100 Books from the Waller Collection in the University Library of Uppsala illustrating the development of Plastic Surgery*, (Uppsala, 1955) .

50 J. Ferguson *Bibliotheca Chemica*, (Glasgow, 1906); D. I . Duveen, *Bibliotheca Alchemica et Chemica*, (London, 1942).

Fig. VII. *The frontispiece of Dante's* Comedia, *(Venice, 1529), bearing the inscription "Joanes baptista a porta amicorumque suorum".* (Bibliotheca Walleriana, *no. 19549.*)

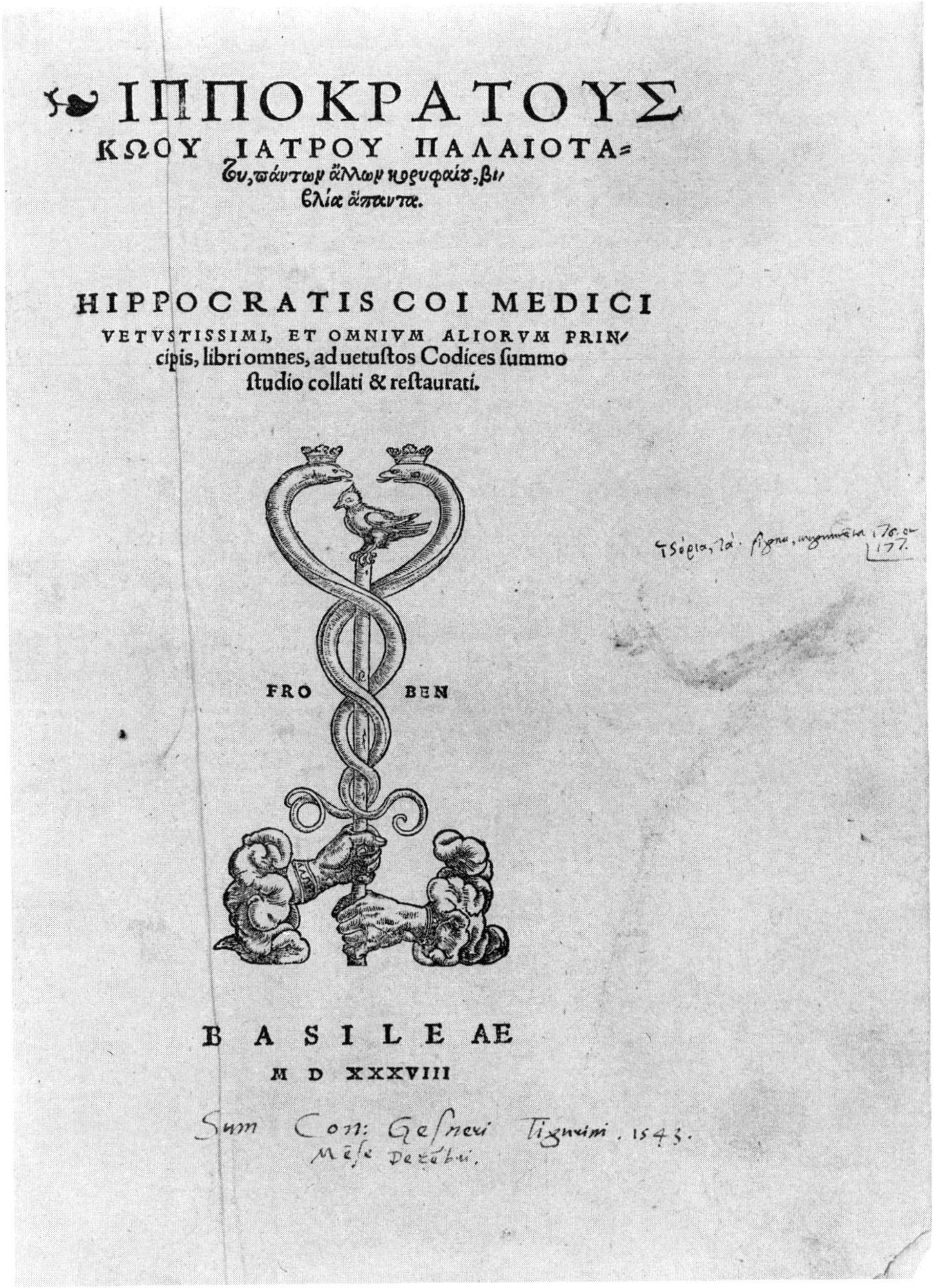

ΙΠΠΟΚΡΑΤΟΥΣ
ΚΩΟΥ ΙΑΤΡΟΥ ΠΑΛΑΙΟΤΑ-
του, πάντων ἄλλων κορυφαίου, βι-
βλία ἅπαντα.

HIPPOCRATIS COI MEDICI
VETVSTISSIMI, ET OMNIVM ALIORVM PRIN-
cipis, libri omnes, ad uetustos Codices summo
studio collati & restaurati.

FRO BEN

BASILEAE
M D XXXVIII

Sum Con: Gesneri Tigurini. 1543.
Mēse Decēbri.

Fig. VIII. *A Froben edition of one of Hippocrates' Works bearing the inscription "Sum Con: Gesneri Tigurini. 1543. Mense Decembri".* (Bibliotheca Walleriana, *no. 4486.*)

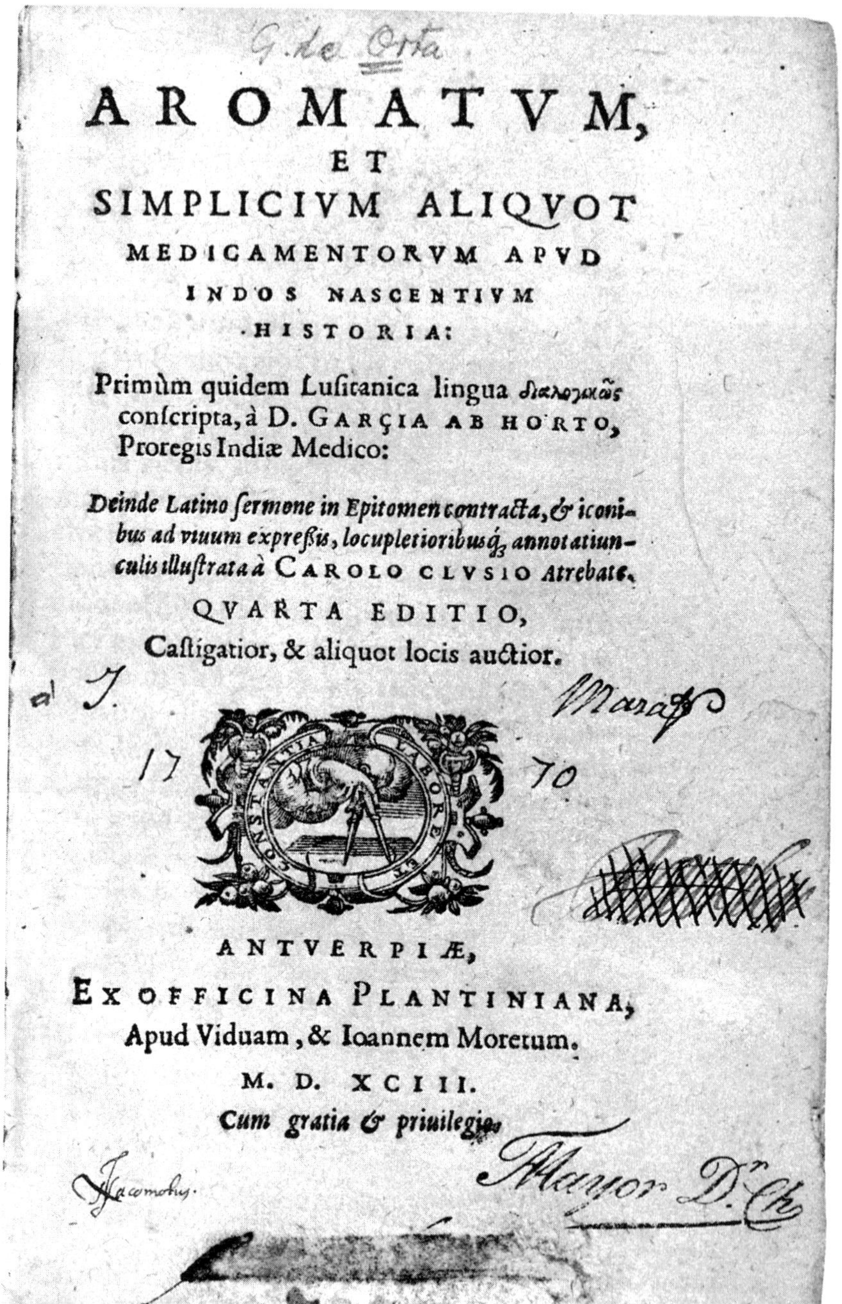
AROMATVM,
ET
SIMPLICIVM ALIQVOT
MEDICAMENTORVM APVD
INDOS NASCENTIVM
HISTORIA:

Primùm quidem Lusitanica lingua διαλογικῶς conscripta, à D. GARÇIA AB HORTO, Proregis Indiæ Medico:

Deinde Latino sermone in Epitomen contracta, & iconibus ad viuum expressis, locupletioribusq; annotatiunculis illustrata à CAROLO CLVSIO *Atrebate.*

QVARTA EDITIO,
Castigatior, & aliquot locis auctior.

ANTVERPIÆ,
EX OFFICINA PLANTINIANA,
Apud Viduam, & Ioannem Moretum.
M. D. XCIII.
Cum gratia & priuilegio.

Fig. IX. *Frontispiece of Garcia de Orta's* Aromatum ... historia, *(Antwerpen, 1593), bearing the inscription "à J. Marat".* (Bibliotheca Walleriana, *no. 7006.*)

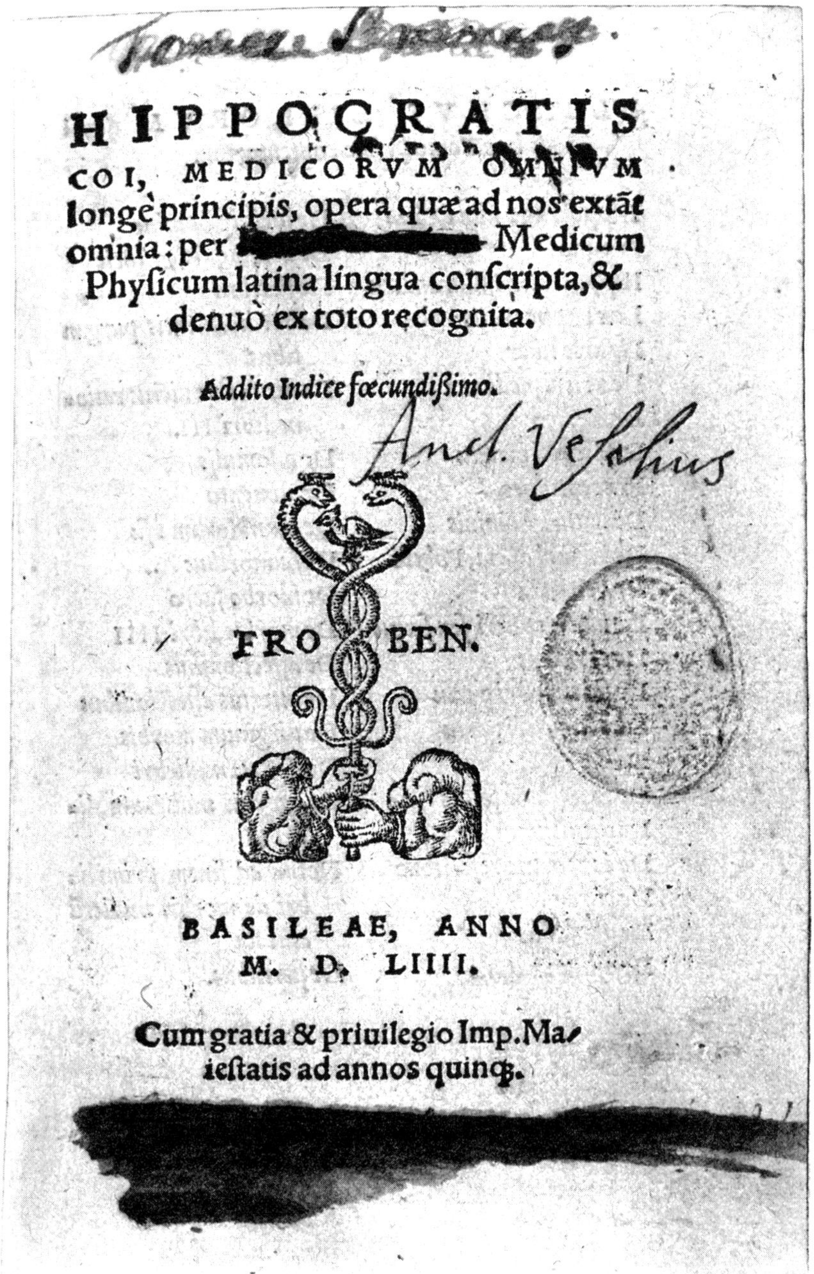
HIPPOCRATIS
COI, MEDICORVM OMNIVM
longè principis, opera quæ ad nos extãt
omnia: per [illegible] Medicum
Physicum latina lingua conscripta, &
denuò ex toto recognita.

Addito Indice fœcundissimo.

And. Vesalius

FRO BEN.

BASILEAE, ANNO
M. D. LIIII.

Cum gratia & priuilegio Imp. Ma-
iestatis ad annos quinque.

Fig. X. *Frontispiece of Hippocrates'* Opera, *(Basel, 1554), bearing the inscription "And. Vesalius" and later inscriptions by Adami Loniceri, William Stevenson and J. Keith.* (Bibliotheca Walleriana, *no. 4498.)*

logue, item No. 18), von Haller, Lavoisier, Linnaeus, Marat (Fig. IX), Moleschott, Oldenburg, Redi, Scaliger, Sennert, Vesalius (Fig. X), Virchow, and many others. These books, which often bear comments and annotations, provide an invaluable and hitherto unexplored source of historical understanding of the relation between scientists and the books they read. A complete catalogue of these books would bring out the importance of the Collection even more.

Regardless of its qualitative composition, if one compares the Waller Collection with those of other prominent collectors of this century, one cannot but notice the astonishing quantitative difference. The medical library of the Canadian physician and collector Sir William Osler (1849-1919)[51] lists "only" 8000 titles. Likewise, the library of the famous American surgeon Harvey Cushing (1869-1939)[52] lists less than one third of the works collected by Waller. In reality it would be impossible to find any private medical library of this century which measures up quantitatively and qualitatively to that of Waller. The only library which had a quantitative similarity to Waller's was collected by the eighteenth century Swiss physiologist and poet, Albrecht von Haller, who put together one of the most outstanding medical libraries of his epoch.[53] Haller's library in fact lists some 9000 titles of books and 11,000 medical and scientific dissertations, a figure which is close to that reported by Sallander in the *Bibliotheca Walleriana*. However, some important differences remain. Waller's library in fact lists titles which are among the rarest works of medical literature of all times, whereas in Haller's library there are only a few works which may be regarded as being of great rarity. Furthermore, Haller's interest in book-collecting was subordinated to scientific concerns rather than bibliophilic ones.

From this brief sketch it emerges quite clearly that the Waller Collection of scientific literature constitutes a unique source for the entire community of historians of science. In this connection, it is significant that, after having received Sallander's Catalogue, the late Professor of History of Medicine at Yale University, John Fulton, wrote a letter to Waller's widow which is worth quoting at length:

---

51 *Bibliotheca Osleriana. A Catalogue of Books illustrating the History of Medicine and Science,* (Montreal, 1969).

52 *The Harvey Cushing Collection of Books and Manuscripts,* (New York, 1943).

53 A catalogue of Hallers's library has been recently published by M.T. Monti, *Catalogo del fondo Haller della Biblioteca Nazionale Braidense di Milano,* 11 vols., (Milan, 1983-1992). The library is now kept in Milan's National Library.

> At long last that wonderful, two-volume catalogue has arrived, and I can't tell you how utterly thrilled I am to receive it. It is not only a most fitting tribute to Dr. Waller and all that he stood for but, like the Osler catalogue, it will always stand as an enduring and inspiring memorial to one of the greatest humanists and bookmen of our time. My only regret is that he could not have seen the finished book. I am sure it would have made him forget the slight disappointment he felt, prior to his death, over the rather unusual classification adopted by Dr. Sallander and the change in the form of entry as well as the omission of his many original notes of provenance and content which would so greatly have enhanced the volume as a scholarly tool.
>
> The first section of his Incunabula most fortunately does retain many, but certainly not all, of Dr. Waller's notes [...] The other three sections on medicine and dentistry in the first volume are admirable, especially for the fact that they include in each instance the indication both of size and pagination. His holdings of those authors on which he had really concentrated, such as William Harvey and Lorenz Heister, are really fabulous. But to me it is a tragedy that his note about the peregrinations of the first edition of the *De motu cordis* was omitted. His collection of Linné is a bibliography in itself, and as with the Harveys, I hope the authorities at Uppsala will one day use the collection as the basis of a definitive bibliography of the author in question, in which case Dr. Waller's notes could be used and amplified [...]
>
> Ancient medicine is wonderfully well represented, especially Galen, with all the early French translations. I suspect that his holdings of Hippocrates rival, if they do not excel those of the British Museum. It seems especially strong in the early Hippocratic translations into French, German, Spanish, and other European languages.
>
> Uppsala will now become a Mecca for students of the history of medicine and science, and I am planning, now that I have become a professor of medical history, to send my students there whenever opportunity permits.[54]

Fulton's great enthusiasm for the Collection which was shared by many of his American as well as European colleagues did not have the hoped-for effect and, with a few exceptions,[55] the Waller Collection fell into oblivion. The Collection was rarely used for systematic research and the bibliographies projected by Fulton are still remote *desiderata*. This is even more surprising if one considers that Sallander's Catalogue has become the standard bibliography of the history of medicine throughout the world.

Regrettably, the oblivion into which the Collection has fallen during the last 40 years has led to some important losses. In spite of the fact that Waller

[54] From Waller's correspondence in Waller 653 B:1. The correspondence is unfortunately not catalogued and there is no apparent order in its arrangement.

[55] Johan Nordström and Sten Lindroth, who met Waller on several occasions, and Michel Foucault seemed to be the only scholars who had a certain acquaintance with the Collection and made partial use of it in their scholarly work. After the death of Kleberg and Sallander, however, the Collection has fallen into oblivion.

Fig. XI. *Contemporary marginalia to Carl Wilhelm Scheele's* Traité chimique de l'air et du feu, *(Paris, 1781). (Courtesy of Ove Hagelin.)* (Bibliotheca Walleriana, *no. 11235.*)

put as a condition of his donation that the books remained within the walls of the University Library, some books have been lent to other European libraries for exhibitions and some duplicates have been given to other Swedish libraries and eventually sold(!) This is for instance the case with a rare French translation of Scheele's *Chemische Abhandlung von der Luft und dem Feuer* (1777), bearing several marginalia of an unknown French chemist[56] (Fig. XI). Other books, fortunately not many, have suffered the same fate.

Thanks to a stricter policy, the situation has considerably improved in recent years but the Waller Collection of books still remains an unexplored and unexploited treasure of learning. The books, however, are only a part, and by no means the most significant, of the Waller Collection. There are, as we shall see, other valuable items, today scattered in different parts of Uppsala University Library, that belonged in fact to Waller.

AUTOGRAPHS AND MANUSCRIPTS

In the preface to Sallander's Catalogue, the late Chief Librarian of Uppsala Universty Library, Tönnes Kleberg, wrote:

> Dr. Waller was active as a collector from the early 1900's. At first manuscripts and autographs were his main interest; he collected nearly 20,000 letters dating from the beginning of the 16th to the early 20th century. A great many of these letters have been written by or addressed to men famous in the history of medicine. Yet this collection of letters goes beyond [these] boundaries.[57]

These lines passed nearly unobserved when they were written and today, after almost 40 years, they have been totally forgotten. It is therefore of a certain interest to describe more in detail the composition and the genealogy of this part of the Collection. As Kleberg rightly claimed, Waller began his career as a bibliophile by collecting autographs and manuscripts connected with the history of science and medicine possibly at the end of the 19th century.[58] In 1953 Waller contacted Tönnes Kleberg and expressed to him

56 I owe this information to the courtesy of Ove Hagelin, who bought the book from the Library of the Karolinska Insitutet in Stockholm. Scheele's work, which is the *Traité chimique de l'Air et du Feu,* (Paris, 1781), still bears the ex-libris of Uppsala University Library: *Bibliotheca Walleriana*, no. 11235.

57 Sallander, *Bibliotheca Walleriana, op. cit.*, vol. 1, ix.

58 The collection of autographs and manuscripts is hardly known but some concise information can be drawn from the studies by Lindroth and Krook quoted in notes 41 and 49.

the wish to sell his collection of manuscripts and letters to Uppsala University Library, so that it might complement his former donation of the books. In May 1954, Waller wrote to Kleberg a letter in which he made the following description of his holdings:

> Dear Tönnes Kleberg,
> In accordance with your wish I here send the catalogues of my autograph collection that I have at present; there are four of them. I made the entries I-III quite some time ago, which is why I cannot guarantee their 100% reliability: over the years it is possible that some autograph has been used in exchange for some book; as a rule it would then have been struck out of the catalogue, but I would like to make the reservation that on some occasions such a correction of the catalogue *may* have been neglected. Catalogue no. IV has recently been completed and is therefore exact in the information it contains. I should also add that in addition to those registered in catalogues I-III there exists a considerable number of autographs that have not been registered; to these belongs a large French collection of autographs exclusively of doctors, most of them with accompanying portraits.
>
> The catalogue entries, even in the state they now exist, ought to provide an overview of the nature and range of the collection.
>
> Concerning the division "Manuscripts" in my list the majority of them were sent in 1951 along with the book collection to Uppsala University Library[;] some: no. 10 Bodoni, 12 Valles and Olivares, 15 Thurneysser, 32a Baglivi and 49 Virchow are still with me but have been set aside for sending when you wish.
>
> In the matter of *diplomas* there are in addition to the 21 registered a further large number of them which have not yet been catalogued, due to the fact that I haven't managed [to put them in order], and concerning the rest because quite simply I haven't had time to enter them. Among these there are certainly many that it would be of great interest to have put into the catalogue. Some of these have already been packed for sending.
>
> The same applies to the alba amicorum.[59]

As we can see from this letter, Waller himself did not have an accurate idea of the number and quality of his Collection. What is sure is that before submitting the offer to Kleberg the famous London bookdealer, E. Weil, made him an offer in 1950 of £2000 for a part of it.

As soon as Kleberg saw the catalogues compiled by Waller, he immediately understood the immense scientific value of the Collection and the importance of ensuring it for Uppsala University Library. Thanks to Kleberg's effort, the Collection was finally purchased for the total amount of 50,000

[59] This letter is kept in (UUB) Bibliotekets arkiv, ÖB:s Korrespondens DC 1947-1955, Erik Waller.

Swedish Crowns,[60] a price which was considerably lower than that offered by Weil some years before. As an addition to the Collection, Waller's widow also donated to the Library the private correspondence of her husband at the request of Kleberg. It is likely that the Waller collection of engravings and medals was acquired together with the autographs. In 1970, after the death of Waller's widow, a portrait of Paracelsus (see Catalogue, item No. 7) and a few other items were donated to the Library.

The Waller collection of autographs was estimated by Kleberg in 1954 to around 20,000 items. This figure, however, was a rough estimate which did not take into account the acquisitions made by Waller during the last years of his life. In reality, a reliable calculation of the number of letters kept in the collection is extremely complex. There are in fact more than 8000 envelopes each of which might contain from one up to 40-50 letters. In addition to this, there are 14 volumes containing autographs (around 3000) and some 2000 autograph letters of mostly French physicians which Waller had no time to put in order or catalogue. From my own calculations, which also are approximate, the collection of autographs and letters exceeds 30,000 items. Obviously this figure has to be taken as a rough estimate of a collection which has never been catalogued or studied in a systematic way.

The collection of autographs and letters is kept in 79 boxes and 14 volumes. Waller himself began classifying the autographs by nationality and within each nation they are usually arranged in alphabetical order. The largest collections of autographs are the German and Austrian ones which are kept in 16 boxes. This is followed by France (13 boxes), Sweden (9), England and America (7), Italy (3), etc. Within this arrangement, which is still useful and constitutes the one used by the Staff of the Manuscript Department, there are many inconsistencies. Accordingly, one might find a letter by the Dutch anatomist Peter Camper both in the box containing autographs of Dutch scientists and in the box of one of the most famous of his addressees, i.e. Linnaeus. Likewise, one might find the letters of the Italian erudite, Ludovico Muratori, both in the box containing the Italian autographs and in the file "Newton" in which one can find some 40 letters addressed by various European scientists to the English physicist. Sometimes the letter of a certain

[60] Today the corresponding amount would be approximately 800,000 Swedish crowns. The acquisition was made thanks to the donation of 25,000 Crowns by the Knut and Alice Wallenberg Foundation, 15,000 Crowns by the King Gustaf VI Adolf 70-Years Fund, and 10,000 Crowns by the Uppsala University Reserve Fund.

author was mistakenly arranged by Waller together with others belonging to another nation. Accordingly, a document written in French signed by the Italian Renaissance engineer, Agostino Ramelli, can be found in the French box because it concerned Ramelli's residence in Paris. Cases like this are not at all unusual and it is in principle impossible to know whether an autograph of a certain author can be found or not in the collection. After all, since Waller himself candidly declared his difficulties in describing the collection, it is not surprising that we are faced with this kind of problem. By contrast, given the dimensions of the collection it is astonishing that the Staff of the Library can still trace most of the documents requested by readers. The handwritten catalogues compiled by Waller provide librarians as well as scholars with the only source of orientation to the collection. There are five numbered catalogues; no. I lists in alphabetical order the autographs of German and Austrian provenance, no. II those from France, no. III. those from Italy, Holland, Belgium, England, USA, and other countries, no. IV lists some detailed bibliographical descriptions of autographs of famous scientists, and no. V offers a list of names of famous physicians in alphabetical order to each of which, possibly, corresponds an autograph or a manuscript in the collection. The first three catalogues often provide the dates and number of the letters written by each author, but their reliability is not very high. First of all, several autographs which have not been cancelled from the catalogue cannot be found; secondly, several thousands of autographs are not listed at all; thirdly, during nearly 40 years of use, a certain number of autographs and manuuscripts (some of great importance) have been misplaced, lost, or stolen. In addition to these problems, autographs of the same author can be found in three and sometimes four different places, in spite of being recorded for only one.

If the quantity of autographs in the Waller Collection is overwhelming, the quality of many of its items make them of great interest for both the historian of science and the historian of ideas. To quote some celebrities at random, we can see that the collection lists letters and autographs by prominent physicans such as Baglivi, Blumenbach, Boerhaave, Cabanis, Fracastoro, Freud, Gall, Guillotin, Hales, Lancisi, Leewenhoek, Marat, Mesmer, Pasteur, Quesnay, Helvetius, Licetus, Camper, Pontanus, Ray, Hunter, F. Fuchs, F. Hoffman, Haeckel, Virchow, Pavlov, Malpighi, Röntgen, Wassermann; by astronomers such as Kepler, Galileo, Newton, Maupertuis, Wargentin, Fontenelle, Laplace, Lalande, Delambre, Herschel, Gauss, Schiaparelli, Flamsteed, G.D. Cassini; by naturalists such as Adanson, Cuvier, Dolomieu,

De Luc, Mendel, Lamarck, Spallanzani, Lyell, Gouan, Redi, Darwin, Needham, Treviranus, Owen, Senebier, Linnaeus, Bonnet, Thunberg; by mathematicians such as J. Bernoulli, Babbage, d'Alembert, Lagrange, Cauchy, Collins, Wallis, Monge, Fermat; by chemists such as Mendeleev, Senac, Berzelius, Lavoisier, Scheele, Priestley, Bergman, M. Curie, Boyle, Davy, Faraday, Kekulé, Liebig; by physicists such as Euler, Galvani, Volta, Rumford, Kelvin, Hooke, Ampère, Arago, Bohr, Einstein, Borelli, Torricelli, Viviani, Musschenbroek, s'Gravesande, Celsius; by philosophers such as Rousseau, Condillac, Condorcet, Nietzsche, Ramus, Bentham, Malthus, Locke, Fichte, Hegel, Schelling, Mendelssohn, C. Wolf, Schleiermacher, Thomasius, Leibniz, Nicole, Arnauld, Bacon, Descartes, Voltaire; by prominent literary, political, or cultural figures such as Louis XIII, Louis XIV, Louis XV, Louis XVI, Napoleon, William Pitt, Strindberg, Sade, Guicciardini, Bembo, Eiffel, Daguerre, La Fontaine, Goethe, Algarotti, Rudbeck, Swedenborg, Erasmus, Zwingli, Kircher, Bodoni, Manuzio, and Panckoucke.

This long list of names, which only offers a subjective quantitative choice from the collection, suffices to provide a notion of its richness. Naturally, as often happens in this kind of collection, the documents relating to famous scientists frequently offer disappointing or unoriginal contents which do not enrich our historical understanding of them. However, in a few cases some of these letters or autographs provide important and original insights into significant aspects of the history of science. The few letters or autographs listed in this catalogue hopefully offer relevant examples of this kind of document but the large majority of them still needs to be explored. Even those documents and letters which are not of immediate historical relevance will fulfil the important function of filling the gaps in the catalogues of the correspondence of many scientists which are now in course of publication. In this respect the collection has already offered a rich source of information.[61]

Among letters and autographs in the Collection, it must be mentioned in passing that a large number of them still need to be identified and attributed. Waller was undoubtedly an expert and accurate historian of science but the importance and the identity of many documents passed unnoticed.

[61] We know that the ongoing editing of the correspondence of Darwin, Blumenbach, Albrecht Von Haller, Kepler, Faraday, Vallisnieri, and Malpighi has benefited from the holdings of the Waller Collection. At the same time it has to be underlined that due to lack of knowledge several editions of correspondence, such as those of Spallanzani, Newton, d'Alembert, Euler, Lavoisier, E. Darwin, a.o., were published without including the letters contained in the Waller Collection.

Another important factor, which adds further scientific value to the Collection, is that in most of the envelopes containing letters and autographs Waller enclosed historical information concerning the author, his scientific contributions, photographs, engravings, etc. Some of this complementary material has sometimes great historical value. In the envelope containing Darwin's letters, for instance, a photograph is enclosed (see Cataloge, item No. 61) bearing his autograph. In other cases, the envelopes contain drawings, leaves of a "Stammbuch" (hereafter Album amicorum), frontispieces of the works with the signature of the author. It is worth noting that in most cases these kinds of autograph documents were not catalogued by Waller in his notebooks. The Collection also contains a good number of medical diplomas from different European universities such as Bologna, Padua, Leiden, Montpellier, Paris, Pisa, Rome, Salerno, Uppsala, Danzig, which are located in three boxes. Some of them, as shown in items No. 13 and No. 32 of this catalogue, are beautifully illuminated and bear the signature of important physicians of the 17th century. Others, no less important from the historical point of view, bear the signatures of J.H. Cuyper, J. Gronovius, A. Vallisnieri, A. Marsigli, and of other famous professors of medicine.

Apart from the diplomas, Waller also collected a considerable number of official documents, statutes, relating to university activities which sometimes bear the signatures of Kings and Heads of State of the 17th and 18th centuries. As there is no trace of them in Waller's handwritten catalogue, these documents need a systematic and thorough investigation.

MANUSCRIPTS

With respect to the manuscripts of the Waller Collection, the situation is no less complex. During the 1970's an attempt to catalogue them was made but the preliminary typewritten catalogue listed only 77 items, that is to say, a catalogue of those manuscripts which were kept separate by Waller from the collection of autographs and letters. This catalogue, the aim of which was to provide a preliminary description and the short-titles of the manuscripts, does not take into account either the manuscripts pasted into printed books, or those of smaller size or unbound which were embodied in the collection of autographs, as well as those in large folio which could not be placed in a box. Although a more complete list of the manuscripts is a work which remains

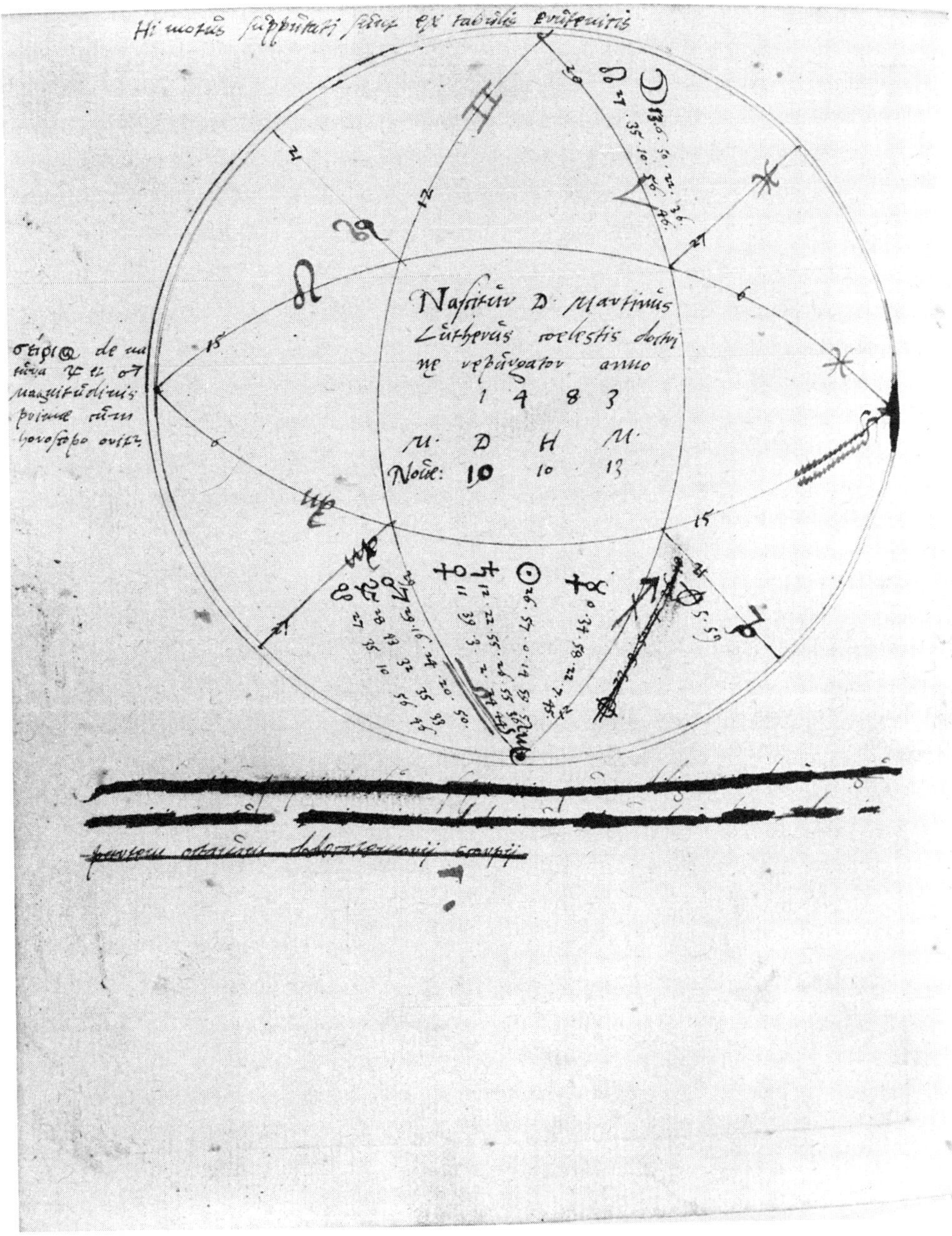

Fig. XII. *A sixteenth century handwritten horoscope of Martin Luther. The manuscript, which contains serveral horoscopes of prominent Renaissance figures such as Vesalius, Melanchthon, Camerarius, Copernicus, is pasted in Abū 'Alī al-Hajjāt's* De judicis nativitatum, *(Noribergai, 1549).* (Bibliotheca Walleriana, *no. 11992.*)

to be done, from our preliminary survey it emerged that the number of manuscripts of the Waller Collection by far exceeds 100. This figure is still approximate because in the collection of autographs we have found dozens of envelopes containing manuscript memoirs, short essays, drafts of lectures, treatises, poems, etc., which obviously must be regarded as manuscripts. Relevant examples of these uncatalogued manuscripts are described in items No. 8, No. 23, No. 59, No. 63, and No. 65 of this catalogue.

If, on the one hand, we cannot at this stage be more precise on the number of manuscripts in the collection, several items provide us with some useful indication as to its quality. In addition to those described in this catalogue, Waller owned some important manuscripts primarily but not exclusively dealing with the history of medicine and science. Among these there are a 15th century version of Gilles de Corbeil's *Liber de urinis;* a Latin poem entitled *Speculum puerorum* dated 1447; a 16th century manuscript containing some 80 horoscopes of notable figures of the Renaissance, such as Luther, Melanchthon, Vesalius, J. Camerarius, Copernicus, a.o. (see Fig. XII); a treatise on the healing of wounds by Leonnhart Bagasser; Petrus Camper's surgical treatise (in Dutch): a contemporary transcription of the anatomical lectures delivered by the British anatomist, Alexander Monro, in the year 1752; a manuscript of the Mexican physician, Joseph Guadalupe Eliseo Ximenes, dated 1762, on the legal and scientific structure of the practice of medicine in Mexico City; an extremely important manuscript of the German physiologist, Caspar Friedrich Wolff, entitled *Grundsätze der Physiologie oder die Wissenschaft von Verrichtungen des menschlichen Körpers,* (Berlin, 1766); the lectures on surgery delivered by Sir Ashley Cooper at St. Thomas Hospital in 1807. The collection also lists a.o., manuscripts of Leonhardt Thurneysser, Nicola Boldone, Antoine and Jean Louis Petit, Adolph Murray, August Gottlieb Richter, and transcriptions of lectures and manuscripts by Aristotle, Servetus, T. Bartholin, T. Sydenham, Linnaeus, D. Cotugno. The list could be extended but the few manuscripts that have been mentioned will hopefully suffice to encourage historians of science to begin a systematic investigation of the most significant of them.

ALBA AMICORUM

"In the hidden recesses of the Library are preserved few documents that on account of their historical flavour and personal stamp can offer greater interest and pleasure than the Alba amicorum. They provide inexhaustible sources for

Fig. XIII. *A leaf of Stolz von Stolzenberg's Album amicorum with the portrait of the Dutch instrument-maker Cornelius Drebbel. (Uppsala University Library, shelfmark Y 132d.)*

research in literary history, the history of ideas and biograhy. Many valuable poems, many important personal references to do with the history of ideas and biography, have, thanks only to them, been saved for posterity".[62]

The Swedish libraries and archives have a rich collection of Alba amicorum and according to Åke Davidsson their total amount is 360. Waller collected 37 of them, more than 10% of a collection which – after the German one – is regarded as among the richest in the world.

The peculiarity of this kind of manuscript and its rarity in countries outside Germany and Scandinavia, deserves a brief introduction. The Renaissance German term "Stammbuch", eventually translated as "Album amicorum" or "Philotheca", has no corresponding word in the English language. The modern term "visitor's book" reflects but vaguely the meaning and the historical function played by the Alba amicorum during the 16th, 17th and 18th centuries.[63] A good general definition of them was given by Åke Davidsson:

---

62 T. Kleberg quoted in Åke Davidsson (ed.), *Stamboken, Poesialbumets och gästbokens föregångare,* (Uppsala, 1981), p. 7.
63 On this, see Davidsson, *op. cit.*, pp. 13-19.

Fig. XIV. *Drawing of the Swiss painter Matthäus Merian. From Stolz von Stolzenberg's Album amicourm. (Uppsala University Library, shelfmark Y 132d.)*

> The notion of Stambok [Album amicorum] can have several meanings, but in the light of these quotations it ought to be clear that in this connection it has to do with commemorative books that travellers took with them on their journeys in earlier times and in which entries were made by the more or less prominent persons that they met.[64]

This tradition began during the 16th century in Germany when the gentry exchanged pictures of their arms with each other. This initial exchange developed into something more systematic during the 1550's when the Alba amicorum became small booklets where a traveller collected arms, mottos, drawings, etc., of the distinguished persons he met.

The oldest Album of this kind is considered to be one that belonged to a student of Wittenberg, Cristoph von Teufenbach, dated 1548.[65] From Germany, where the Album had a successful tradition among university students and scholars during the 17th and 18th centuries, it expanded into Switzerland, Holland, and into the Scandinavian countries. It is not surprising then to see that almost all the surviving Alba kept in Uppsala University

64 *Ibid.*, p. 13.
65 *Ibid.*, p. 14.

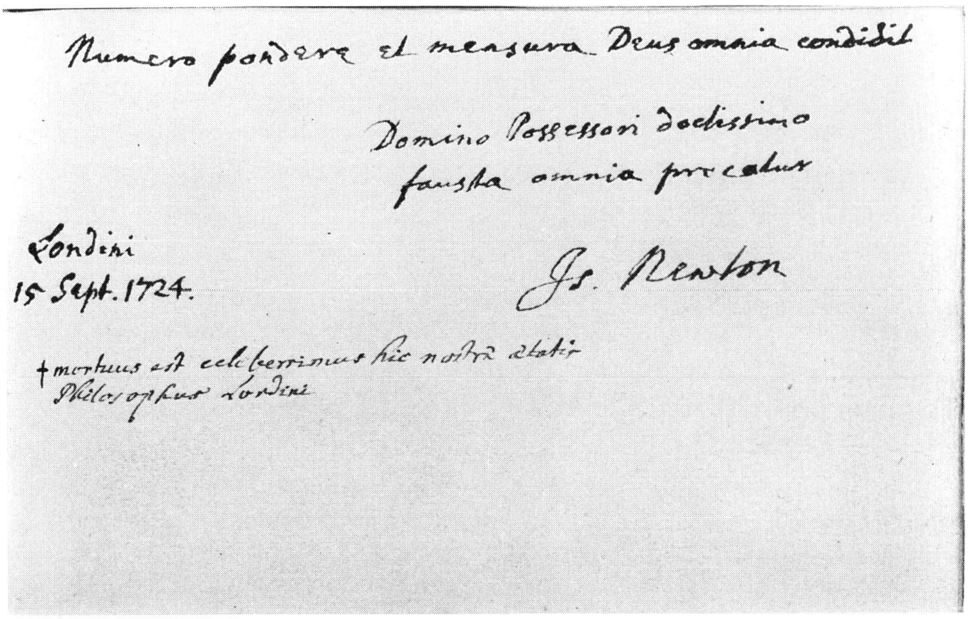

Numero pondere et mensura Deus omnia condidit

Domino Possessori doctissimo
fausta omnia precatur

Londini
15 Sept. 1724.

Is. Newton

† mortuus est celeberrimus hic nostri aetatis
Philosophus Londini

Fig. XV. *A leaf of C.L. Liptenius' Album amicorum bearing Newton's autograph and dictum. (Uppsala University Library, shelfmark, Y 82b.)*

Library (c. 130) come from these countries. The 37 Alba owned by Waller date from the 16th to the 18th centuries and contain more than 4000 autographs. As in the other parts of the Waller collection of Alba amicorum, it is not difficult to find some remarkable items. An interesting example is the Album belonging to Daniel Stolz von Stolzenberg, physician, poet, and alchemist, living during the first half of the 17th century in Kronstadt, a small village in Bohemia. This Album is particularly beautiful because it is adorned with 100 copper etchings of Stolz' work *Chymisches Lustgärtlein* (Frankfurt, 1624). The Album contains some 200 autographs collected during Stolz' peregrinations through Europe in 1622-1627.[66] Apart from several leaves containing the autographs of some relatively famous German alchemists and printers, the Stolz' Album contains leaves signed by the English alchemist Robert Fludd (see Fig. IV), the Dutch cartographer Willem Blaeu, the naturalist Caspar Bauhin, the physicians S. Santorio and F. Licetus, M. Ruland, L. Osiander, Etienne de Clave, a.o. The leaf signed by the Dutch instrument-maker, Cornelius Drebbel, is accompained by a beautiful self-portrait (see Fig. XIII). Another rarity of this Album is the drawing of a

[66] Stolz' Album has the shelfmark Y 132d.

landscape, probably outside Basel, made by Matthäus Merian (1593-1650), famous Swiss painter, illustrator, and engraver (see Fig. XIV).

In the Album belonging to the German theologian Christian Ludwig Liptenius (1698-1759), which he brought with him on his long journey in Europe in the 1720's there are 125 inscriptions, some of which are of great historical interest. Among them we can find the autograph made by Newton when he was 82 accompanied by the motto *numero pondere et mensura Deus omnia condidit* (see Fig. XV). Other relevant leaves of this Album are those signed by Richard Bentley, Musschenbroek, Hartsoeker, Scheuchzer, Ruysch, Christian Wolf, Thomasius, a.o.[67] In the other Alba owned by Waller one can find autographs by Wallis, Oldenburg, Scaliger, Kircher, J. Gruterus, Zwingli, Linnaeus, Celsius, T. Bergman, a.o. As these autographs are not catalogued, a fuller investigation of the contents of these Alba would enrich our knowledge of the lives of scientists with new and original information concerning their contacts.

ENGRAVINGS AND EX-LIBRIS

Waller owned a remarkably rich collection of engravings, caricatures and other iconographical material dealing with the most prominent figures in the history of science. The collection is kept in 22 folio boxes containing some 3000 iconographic items illustrating the history of science. There are woodcuts, copper etchings, engravings, watercolors, drawings, and photographs. Most of them are portraits of single scientists and in some cases they bear the autograph of the possessor, as shown in the case of the 17th century engraved portrait of Descartes, bearing the signature of the English chemist Robert Boyle (see Catalogue, item No. 17). Some of the engraved portraits constitute unique or extremely rare iconographic items. The collection of caricatures is not particularly rich but contains some beautiful examples of 18th century English and French caricatures of physicians (see Fig. XVI). This collection has been catalogued and is kept in the Maps and Prints Department.

Waller owned a collection of ex-libris from the private libraries of physicians. The collection is not very large and primarily lists items from our century but there are some beautiful examples from 18th century medical libraries (see Fig. XVIII).

---

67 Liptenius' Album has the shelfmark Y 82 b.

Fig. XVI. *An eighteenth century French caricature satirizing the doctrine of animal magnetism.*

Fig. XVII. *An engraved portrait of the Italian anatomist Paolo Mascagni.*

Fig. XVIII. *An ex-libris from the library of the French physician C.E. Horeau, editor of Corvisart's posthumous work* Essai sur les maladies et les lésions organiques du coeur et des gros vaisseaux, *(Paris, 1806).* (Bibliotheca Walleriana, *nos. 2149-2151.*)

MEDALS

Waller owned a rich collection of medals (around 600) dating from the 17th to the 20th centuries. Most of the medals commemorate famous scientists and the items No. 9, No. 30, No. 38, No. 42, and No. 45 are typical examples indicating the quality of the collection. Fortunately, unlike the other parts of the collection, we are in possession of a good and reliable handwritten catalogue of the medals, compiled by Waller himself. The collection is now kept in the *Myntkabinettet* in the main building of Uppsala University.

QUANTITATIVE OVERVIEW

We have already stated in the first part of this introduction that a precise quantitative inventory of the items composing the Waller Collection cannot at this stage be provided. Only a systematic catalogue of the collection would provide us with both its quantitative and qualitative richness. However, keeping in mind the limitation of our survey, the data we have collected provides the following figures:

Autographs and letters	30,000
Manuscripts and Diplomas	300
Iconographic items	3000
Ex-libris	200
Medals	600
Uncatalogued pamphlets, booklets, offprints	4000

In addition to this, mention must be made of one big box containing Waller's correspondence with historians of science, physicians and bookdealers.

THE CATALOGUES

Waller was a very meticulous collector and he not only tried to keep a record of what he acquired by compiling handwritten catalogues but he also carried out original and valuable research on the rarest items he purchased. In this respect, it is worth noting that Sallander's Catalogue was largely based on a two-volume manuscript catalogue compiled by Waller himself. Waller's version of the catalogue is still valuable because in it one can find bibliographical information otherwise impossible to acquire.[68] Waller's catalogues of his collection of autographs and letters are, as we said, useful but unreliable and largely incomplete.[69] Some useful historical and bibliographical annotations made by Waller on his collection of manuscripts, diplomas and Alba amicorum can be found in a box with the shelfmark Waller 653 B:2.The general conditions in which the collection has rested for 40 years have increasingly given rise to the need for a systematic catalogue. In order to meet this need and to evaluate the technical as well as financial difficulties involved, an informal committee which includes the Head of the University Library, two members of the Staff of the Manuscript Department, and some historians of science, has been established to examine the possibility of producing a general catalogue of the collection.

The ultimate aim of such a long-term project would be to provide scholars with a printed catalogue of all the autographs and manuscripts which will fulfil the standards required of a modern bibliography. The variety and complexity of items suggest that such a project will include the simultaneous cataloguing of the Alba amicorum, the medals, the offprints, the ex-libris, the engravings with autographs, the manuscripts pasted in books, and the other parts still to be explored. From our survey in fact, it emerges quite clearly that a cataloguing of manuscripts and autographs would be inevitably incomplete if the book and engraving collections were not re-examined carefully.

The realization of such a project will first of all provide the international community of historians of science with an extremely rich and hitherto little known source of information; secondly it will prevent further loss of invaluable material[70] and will give an exact idea of its quantity and of the method of ordering it. The merit of such a catalogue will be to reveal the personality of Waller as a collector as well as an historian of scientific literature. It will, in addition, constitute a fitting tribute to him for having made the extraordinary gift of his Collection to the University Library of Uppsala.

---

68 Waller's catalogue is available in the Manuscript Department *(Handskriftsavdelningen)* of Uppsala University Library.

69 Waller's handwritten catalogues are also available in the Manuscript Department.

70 Apart from the above-mentioned duplicates which have been sold by the Library of the Karolinska Institutet, several autographs such as those of Mendel, Luther, and a manuscript by Einstein have been misplaced, lost or stolen.

Catalogue

EDITORIAL NOTE

The items are arranged in chronological order, although not too strictly. Transcriptions from manuscripts and letters are with a few exceptions faithful to the original.

1. Old-Style Julian dates and other systems of dating that deviate from the Gregorian calendar have been given as they stand, the only exception being the two Arabic manuscripts.
2. Capitals have been retained only in cases of absolute certainty.
3. Some changes have been made in the punctuation and, with a few exceptions, they have been silently embodied in the text..
4. In general abbreviations have been retained as they stand in the text. However, medieval and Renaissance contractions and abbreviations of words have been expanded.
5. When words have been found illegible or when they are irrecoverable because of damage of the letter or manuscript, they are indicated with the sign [...].
6. The numeration of the illustrations follows that of the Catalogue. In the few cases when two illustrations refer to the same item, they have been distinguished by an added letter: e.g. 18a-18b.
7. The order of the illustrations follows, with a few exceptions, that of the Catalogue.
8. When a document is not autograph, the name of the author has been put in square brackets.

LIST OF ABBREVIATIONS

DSB: C.C. Gillispie (ed.), *Dictionary of Scientific Biography,* 16 vols., (New York, 1970-1980).

Sarton's *Introduction:* George Sarton, *Introduction to the History of Science,* 3 vols., (Baltimore, 1927-1948).

Bibliotheca Walleriana: Hans Sallander, *Bibliotheca Walleriana*, 2 vols., (Uppsala, 1955).

1.

[RHAZES, ABŪ BAKR MUḤUMMAD IBN ZAKARĪYĀ AR-RĀZĪ (CA. 854-935)]

MANUSCRIPT Arabic. Unpublished.

TITLE [An abridgment of the "Kitab al-Hāwi". The Ms. also contains an abridgment of Abū Alī Muskūya's "Al-Ashriba".]

DATE 1255 (622 A.H.).

DIMENSIONS 24.6 x 16 cms.

FOLIATION 197 leaves.

BINDING Ornamental black morocco. XVIth century.

SHELFMARK Waller 653 D:l.

HISTORICAL NOTE

Physician, physicist and alchemist, Rhazes wrote among other things a medical encyclopedia (*Kitab al-Hāwi*) which was based on several Greek and Hindu medical sources. According to Sarton (*Introduction,* vol. 1, p. 609) "The al-hāwi has not been published, and there is not even a single complete manuscript in existence [...]. A Latin translation, *Liber dictus Elhavi*, appeared in Brescia (1486), followed by various Venetian editions". The second Ms. is an abridgement of Al-Ashriba, a work on drugs and medicaments. Both abridgments were written by Ibn at-Tilmīdh al-Baghdādī (died c. 1165), a physician who lived and practised at the Court of the Caliph of Bagdad.

2.

[JOHANNES SACROBOSCO (DIED 1256)]

MANUSCRIPT Latin on vellum.

TITLE [Tractatus de Sphaera mundi.]

DATE Second half of the XIIIth century. Although the exact date of the manuscript remains uncertain, the handwriting suggests that it belonged to the second half of the XIIIth century. The binding, made out of a leaf of a Latin manuscript, provides a clearer indication of the approximate date of the Ms. The interior part of the binding facing the first leaf is in fact dated October 1277.

DIMENSIONS 17.5 x 11.4 cms.

FOLIATION 16 leaves written in two columns; the Ms. bears a Latin com-

بسم الله الرحمن الرحيم
الحمد لله حق حمده وصلواته على سيدنا محمد خير خلقه وعلى اله وصحبه
مختار من كتاب ابي بكر محمد بن زكريا الرازي المعروف
بالحاوي
اختاره هبة الله بن صاعد بن ابرهيم المعروف بابن التلميذ وبعضه مصلح
له صفة حب مارستاني نافع من العلل البلغمية القوية
في الدماغ جندبادستر نصف درهم شحم الحنظل ربع درهم
فربيون دانق ايارج فيقرا درهم وهي شربة … صفة دوا
نافع من الرعشة في اليدين لا غاية له تطبخ الرطبة
اعني الاسفست ويدق حتى تصير مثل المرهم وتضمد بها اليدين كل
يوم مرتين فانه يبرؤه البتة ٥ زعموا ومما ينفع من
استرخا الاعضا ان تطبخ الفوة طبخا جيدا شديدا وتصب على
الاعضا حب مرتضى جيد للفالج ايارج عشرة دراهم
قنطوريون دقيق ثلاثة دراهم شحم حنظل درهمين جندبادستر درهم
الشربة وزن درهمين قال الجندبادستر نافع للرعشة شربا ومسوحا
ولجميع علل العصب الرطبة وان كان هناك حمى
لا تستعمله مع حمى قوية ولكن قد دل على قلة خطره دوا

Fig. 1.

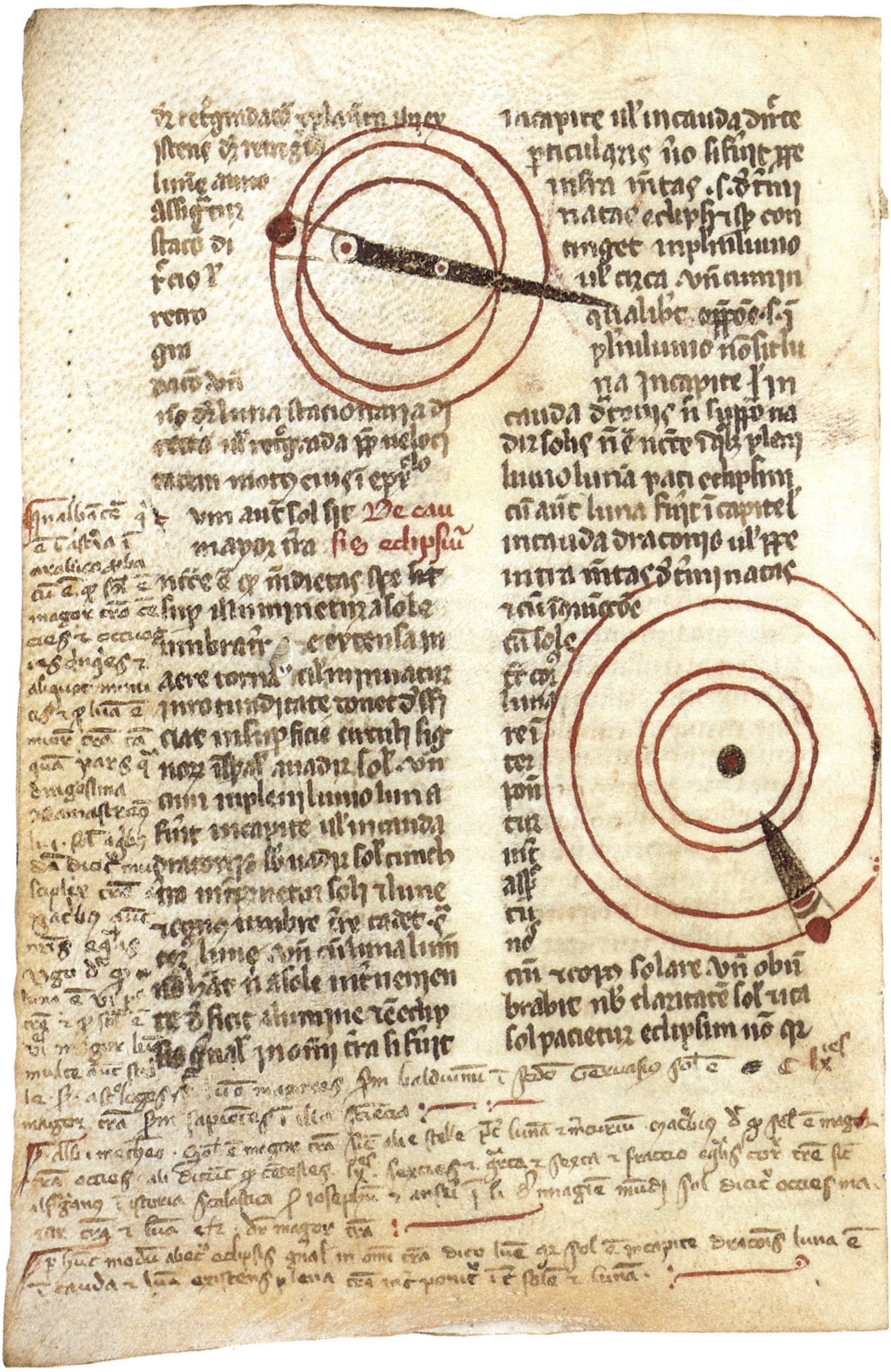

Fig. 2.

	mentary (approx. d. XIVth century) on the margins of the text and on leaves 14-16.
ILLUSTRATIONS	8.
INCIPIT	Tractatum de Spera mundi iiii capitulis distinguimus dicentes primo quid sit opera.
BINDING	Cover of vellum.
SHELFMARK	Waller 653 E:1.
HISTORICAL NOTE	

The *Tractatus de sphaera* (c. 1233) was one of the most important and successful astronomical works of the late Middle Ages and early Renaissance. It was used as a university textbook in order to provide students with a simple but up-to-date treatise on the nature and shape of the earth, of the heavens, the movements of the celestial spheres, the orbits of planets, the rising and setting of the stars, the climates, the movements of the sun, the eclipses, etc. The rarity of this manuscript, which was bought by Waller from a French bookdealer, derives from its early date of composition which makes it nearly contemporary to Sacrobosco's original work.

3.

AMULĪ, SHAMSADDĪN MUḤAMMAD IBN MAḤMŪD

MANUSCRIPT	Arabic. Unpublished.
TITLE	[Sharh Al-Qānūn (Commentary on Avicenna's Canon of Medicine).]
DATE	1359 (760 A.H.)
DIMENSIONS	23.6 x 19.3 cms.
FOLIATION	336 leaves.
ILLUSTRATED	
BINDING	Black morocco, c. XVIIth century.
SHELFMARK	Waller 653 D:l.
HISTORICAL NOTE	

Arabic commentary on Avicenna's Canon (see item No. 6) made by Amulī Sahmsaddīn who was not known to Sarton. The Ms. contains some interesting geometrical illustrations related to the anatomical description of the human body and to anatomy. The triangle shown in the lower part of the leaf of Fig. 3 provides a description of the bones of the human nose and of the jaw. The representation of anatomical objects through geometrical figures reveals the difficulty of perceiving the human body in a realistic form and the attempt to submit scientific iconography to formal and abstract features. Interestingly this illustration corresponds to one preserved in a manuscript copy of Avicenna's Canon kept in Leiden library described by P. De Koning in his *Trois Traités d'Anatomie Arabes,* (Leiden, 1903), p. 463.

4.

LIBELLUS DE MEDICINA

MANUSCRIPT Latin, on vellum. Unpublished.

TITLE Libellus de medicina. Manuscript coming from the Chartreuse of Buxheim with the inscription on the first leaf "Cartusia in Buxheim".

DATE End of the XIVth century.

DIMENSIONS 16.6 x 11.4 cms.

FOLIATION 18 leaves.

INCIPIT "Incipit Libellus de medicina animae."

BINDING Original binding of wooden boards, half covered with pigskin. Original claps.

SHELFMARK Waller 653 E:l.

HISTORICAL NOTE

This manuscript contains a compilation on general medicine and it is divided into the following 12 chapters:

1. De quatuor elementis.
2. De rheumate.
3. De sanguine.
4. De colera rubea.
5. De colera nigra.
6. De commixtione elementorum (which deals with the nature of metals).
7. De quatuor virtutibus humanae complexionis.
8. De medico et novem modis signorum.
9. De dolore capitis.
10. De dolore ventris.
11. De casu capillorum.
12. De dolore frontis.

Fig. 3.

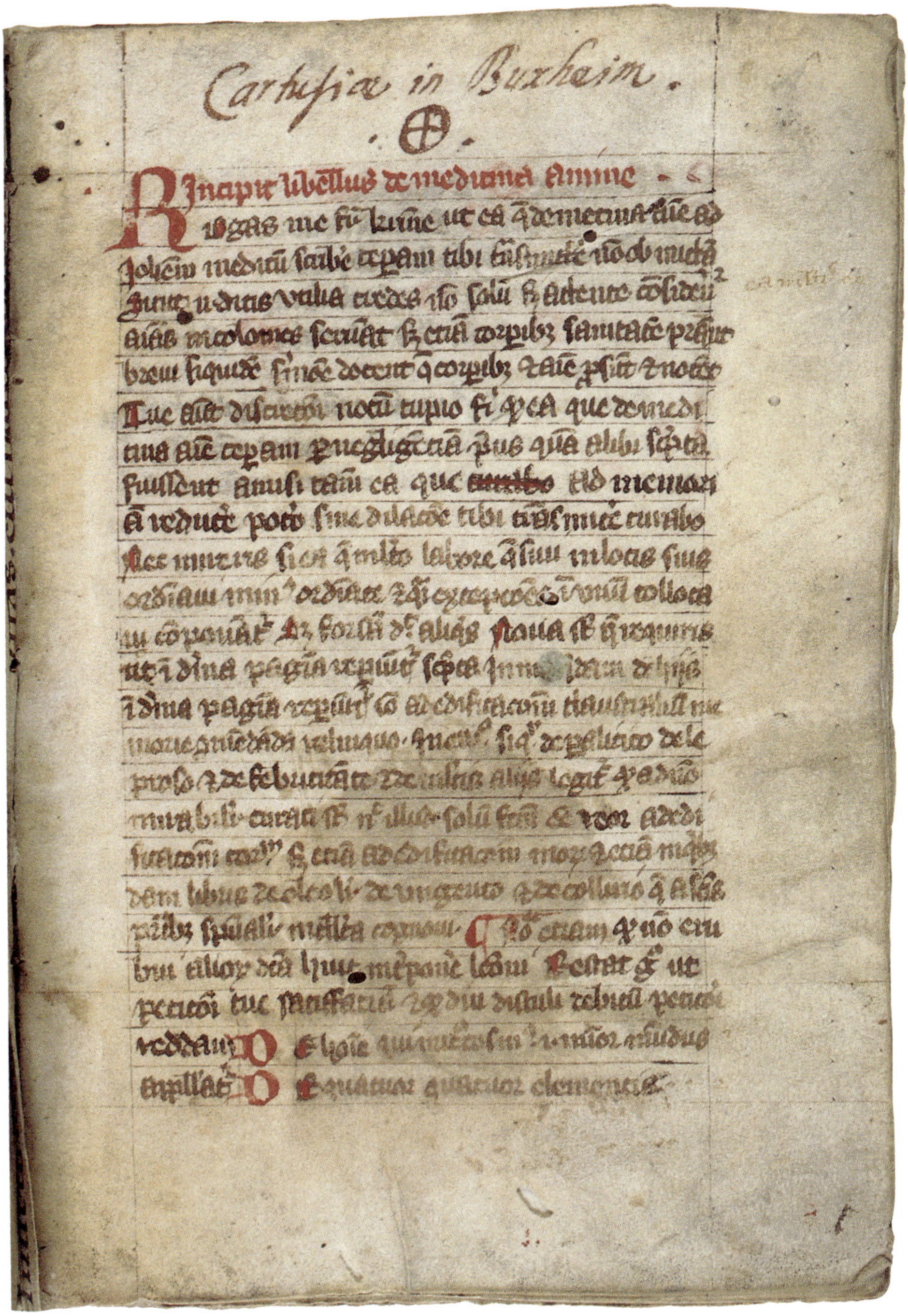

Fig. 4.

5.

[LANFRANCUS MEDIOLANENSIS (SECOND HALF OF THE XIIITH CENTURY BEFORE 1306)]

MANUSCRIPT: Old Dutch. In two columns: chapter headings in red. Unpublished. In the 1960's a critical edition was in preparation by Jan Fredriksen. Dr. Jan Deschamps (Handschriftenkabinet – Kon. Bibliotheek Albert I, Bruxelles) has been working on an edition of the manuscript and has published an essay on it under the title "De middelnederlandse vertalingen van de <<Chirurgia magna>> van Lanfranc van Milaan" in *Koninklijke Academie voor Nederlandse taal – en letterkunde. Verslagen en mededelingen*, 1987:3, pp. 469-472.

TITLE: Van Erzeney. Boeck van Serurgien [Chirurgia Magna].

DATE: XVth Century.

DIMENSIONS: 28 x 20.7 cms.

FOLIATION: 77 leaves.

ILLUSTRATED

BINDING: XVIth century binding of limp vellum.

SHELFMARK: Waller 653 D:3.

HISTORICAL NOTE

Lanfrancus's *Chirurgia magna* (1296) was an important work on the treatment of wounds and fractures. According to Sarton *(Introduction*, vol. 2:2, pp. 1080-81) "various early translations extended Lanfranchi's influence outside of the Latin world [...] The Chirurgia Magna was twice translated into English before the middle of the fifteeenth century [...] A French translation was made by a surgeon of Lyon, William Ivoire (or Yvoire), c. 1490". This Dutch translation seems to be the only one ever attempted in this language. Interestingly, on leaf 24 recto two illustrations of surgical instruments (a sticker and a surgical knife) stand for their corresponding words. This substitution of illustrations for words testifies to the difficulty faced by Renaissance scholars in translating into the vernacular the highly sophisticated technical nomenclature of the classical sources. In this specific case it seems likely that the translator did not find the corresponding Dutch surgical terms for the Latin ones describing the two instruments.

6.

[AVICENNA (IBN SĪNĀ) (980-1037)]

MANUSCRIPT Italian. Ornamental headings and initials in red. Unpublished.
TITLE [Canon – abridgement.]
DATE October 10, 1478.
DIMENSIONS 21.5 x 14.5 cms.
FOLIATION 67 leaves.
INCIPIT I have replaced the contractions with the full words: "Questo sie uno fioreto trato de Auicena per queli che non poleno aver molti libri ..." Last leaf "Finito el fioreto trato de Auicena 1478·a-di·8·Octubre." The manuscript is written in Italian but the spelling of certain words and the construction of the sentences show that it was written by a scholar coming from the Venetian area.
BINDING Modern vellum.
SHELFMARK Waller 653 D:3.
HISTORICAL NOTE

Avicenna, who was probably the most important scientist of the medieval Arab culture, is particularly famous for his medical encyclopedia *Qānūn* (Canon). In this work many contagious, psychological and sexual diseases were described for the first time in the history of medicine.

According to Sarton *(Introduction,* vol. 1, p. 711) "the Qānūn was translated into Latin by Gherardo Cremonese [during the XIIth century] and many partial editions of this translation were printed before 1500". Due to the scarcity of historical evidence related to it, the diffusion of Avicenna's Canon in the vernacular is still rather obscure (on this see N. Siraisi, *Avicenna in Renaissance's Italy,* Princeton, 1987, pp. 46-48). This manuscript, which is a compendia of Avicenna's work, testifies to the early presence of medical research outside the university. Its contents, in fact, are clearly directed to practical therapeutics and addressed to a wider readership; the use of the vernacular supports the hypothesis that the manuscript was composed outside the university. Furthermore, as is stated in the Incipit of the Ms., the work was addressed to people who could not afford many books. This might explain the relatively small size of the book, its length and its popular contents.

In spite of these characteristics the anonymous compiler of the Ms. set up a very ambitious plan and declared: "in questo libreto

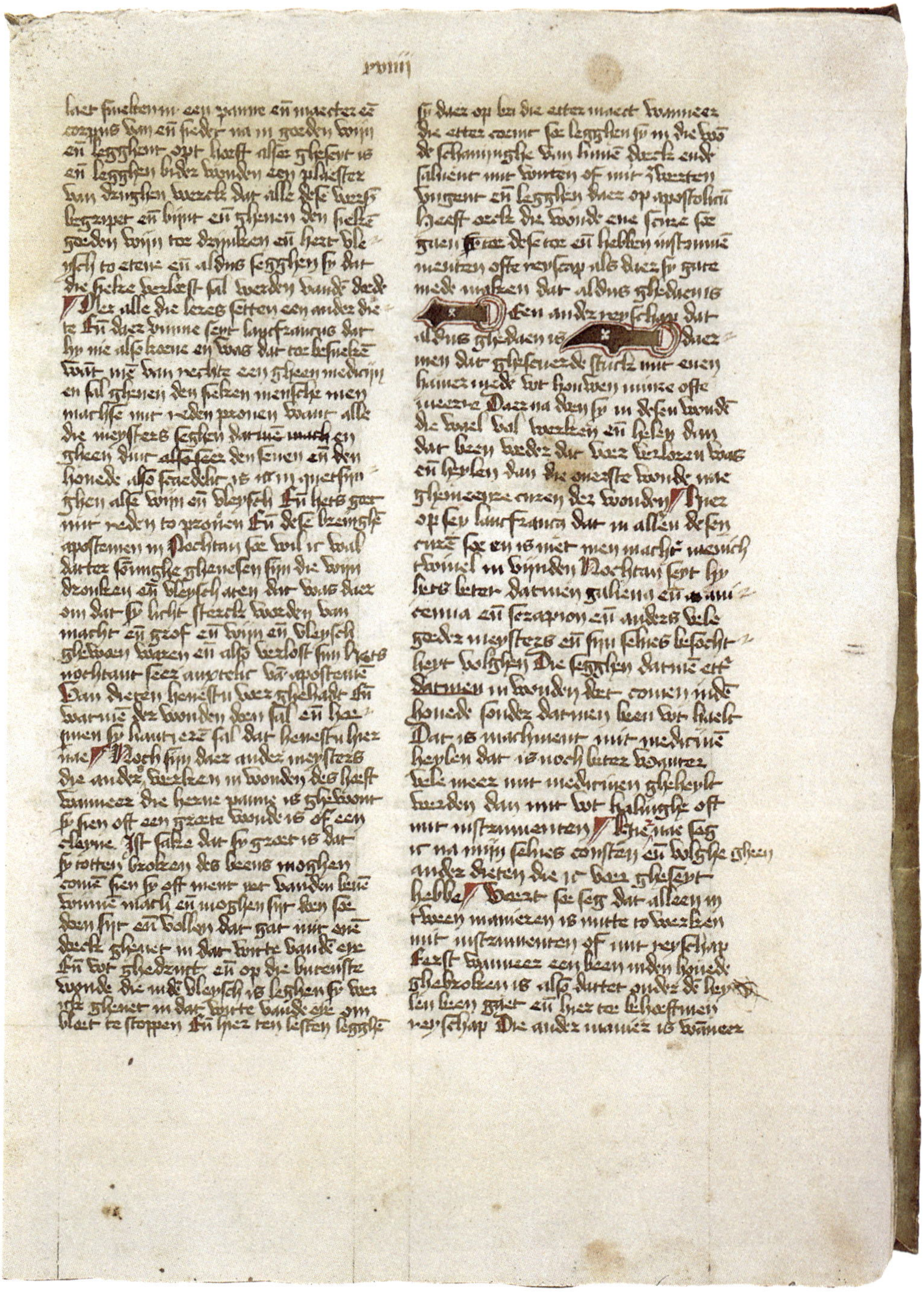

Fig. 5.

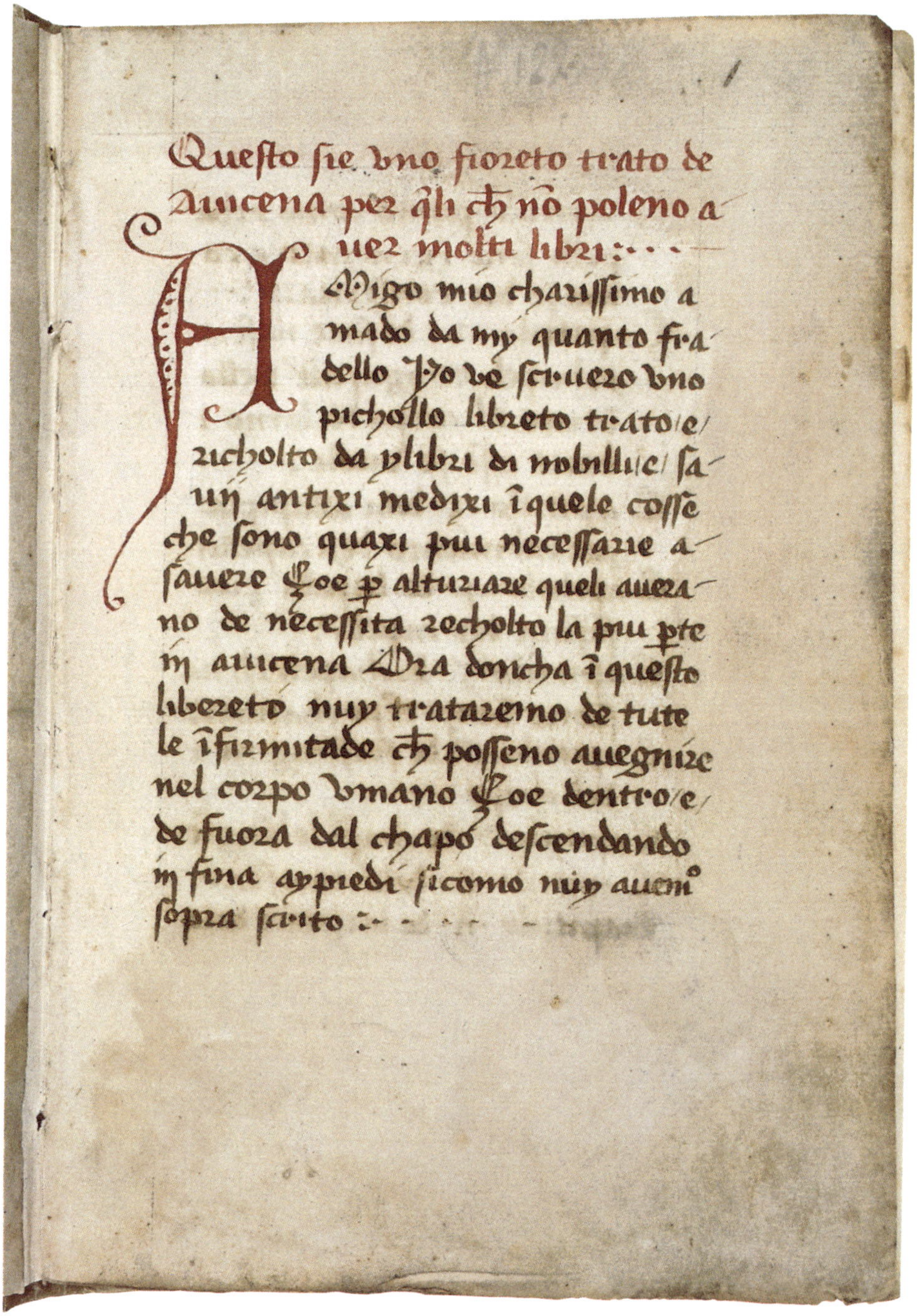
Questo sie vno fioreto trato de
auicena per qli ch nō poleno a
uer molti libri:-..

Amigo mio charissimo a
mado da my quanto fra
dello Jo ve scriuero vno
pichollo libreto trato e
recholto da y libri di nobilli e sa
uij antixi medixi ī quele cosse
che sono quaxi piu necessarie a
sauere Çoe p aiturare queli auera
no de necessita recholto la piu pte
in auicena Ora doncha ī questo
libereto nuy trataremo de tute
le īfirmitade ch posseno auegnire
nel corpo vmano Çoe dentro e
de fuora dal chapo descendando
in fina appiedi sicomo nuy auemo
sopra scrito :- - - -

Fig. 6.

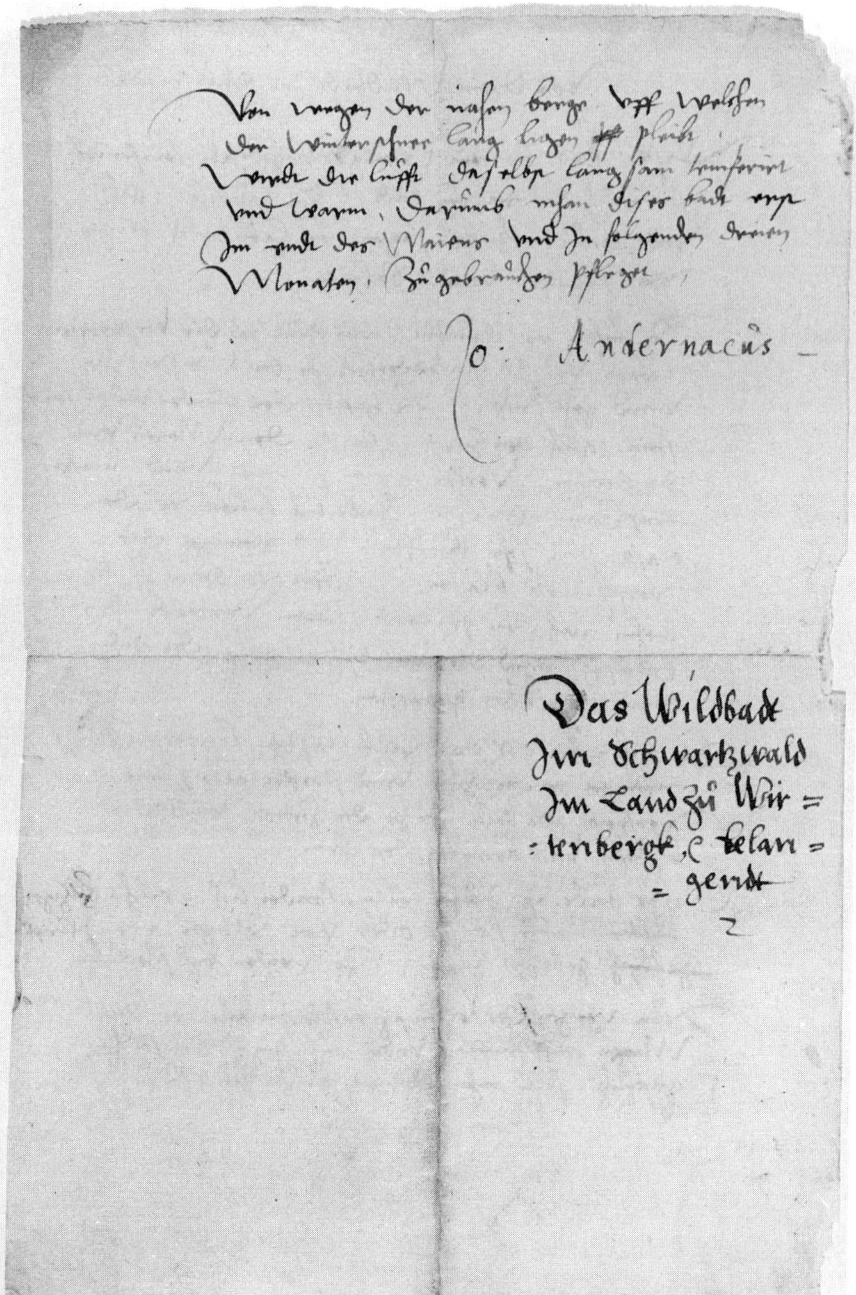
Von wegen der nahen berge uff welchen der Winterschnee lang ligen pleibt, werde der lufft daselbs langsam temperiert und warm, Darumb man dieser badt erst im ende des Maiens und in folgenden dreien Monaten zu gebrauchen pflegt,

Jo. Andernacus

Das Wildbadt Im Schwartzwald Im Land zu Wir= =tenbergk, gelan= =gendt

Fig. 8.

nuy trateremo de tute le infirmitade che posseno avegnire nel corpo umano".

The Ms. is divided into 47 chapters each devoted to a particular disease. Among these diseases, those related to the skin, eye, blood, stomach, ears, liver and other internal organs, are briefly described.

7.

[PARACELSUS, PHILIPPUS AUREOLUS THEOPHRASTUS BOMBASTUS (1493-1541)]

PORTRAIT — Oil painting on wood.

DATE & AUTHORSHIP — First half of the XVIIth century. Flemish school. The back of the painting bears the inscription "Rubens" but this attribution is likely to refer to Rubens' atelier rather than to Rubens himself.

DIMENSIONS — 63 x 50 cms.

LOCATION — Uppsala University Library. Chief Librarian's Office.

HISTORICAL NOTE

Copy of the painting ascribed to Jan van Scorel (1495-1562) and now kept at the Louvre. The original portrait, which shows in the background Dinant, the river Maas, and the bay and rock, is described by Karl Sudhoff in Paracelsus, *Sämtliche Werke*, vol. 6, (München, 1922), p. 16. This copy is said to be of the School of Rubens but no written evidence authenticates this claim. The painting differs from that by Scorel in the depiction of the background landscape and in the depiction of details of Paracelsus' hat and dress. Below the portrait there is the following inscription: "FAMOSO. DOCTOR. PARASEL".

Fig. 9.

Fig. 7.

8.

ANDERNACH, JOHANN GÜNTHER VON (1505-1574)

MANUSCRIPT German.
TITLE Von dem Wildbadt zur Schwartzwald.
UNDATED
DIMENSIONS 39.5 x 20 cms.
FOLIATION 1 leaf.
SHELFMARK Waller Delikatesskaps 2.
HISTORICAL NOTE

Anatomist of the Galenic school, Andernach taught anatomy in Paris to Vesalius and Servetus. His main work *De medicina veteri et nova tum cognoscenda, tum faciunda commentarij duo* was published in Basel 1571. In this work Andernach combined his humanist approach to medicine with the Paracelsian principles. This short Ms. seems to have been written under the influence of Paracelsus; it is in fact devoted to the examination of the therapeutic properties of the waters in Schwartzwald.

9.

[FERNEL, JEAN (1497-1558) AND PARÉ, AMBROISE (1510-1590)]

COMMEMORATIVE MEDAL Described by Horatio Robinson Storer, *Medicina in Nummis. A Descriptive List of the Coins-Medals-Jetons relating to Medicine, Surgery and the Allied Sciences*, (Boston, 1931) n. 1020:

"Ob. JEAN FERNEL. AMBROISE PARÉ. Two busts to right, one of them with cap. EX. LA MEDICINE RENDUE A SON/UNITÉ PRIMITIVE. DÉCRET DU 14 FRIMAIRE/AN III DE LA R.F. (4 dec. 1794.) Signed GATTEAUX.

Rev. AEDES. ACADEMI. ET SCHO. CHIRVRGO. (The first and last letters of the inscription touch the roof. An open court surrounded on three sides the École de Medicine. EX. REGIA MUNIFICENTIA INCHOAT/MDCCLXX ABSOL:/ MDCCLXXIV. signed, N. GATTEAUX. F."

DIAMETER 60 mms.
MATERIAL Bronze.
SHELFMARK Waller Collection of medals, no. 180.

10.

NOSTRADAMUS, MICHEL (1503-1566)

LETTER	French. Unpublished.
PLACE & DATE	Salon de Craux Provence, 25.2.1566.
ADDRESSEE	M. de Saffre.
DIMENSIONS	31 x 21.3 cms.
FOLIATION	2 leaves.
SHELFMARK	Waller Delikatesskaps 2.
TRANSCRIPTION	

Monsieur, i'ay receu vos lettres ensemble les deux escus que m'avez envoyé. Quant à la maladie de Madame [vostre] mere, combien qu'elle ne m'ait rein envoyé cenonobstant ie luy en escris bien au long et de tour son fait. Aussj selon l'exigence de sa maladie ie luy ay ordonné le plus souvreain remede qui se pourroit faire au monde, laquelle donnerez ordre luy faire faire. et le plus promptement que faire se pourra que i'espere en Dieu qu'elle s'en trouvera bien [.] Toutefois sa maladie est fort dangereuse comme ie luy escris. Environ le XVe de jun [pro]chain si Dieu luy fait la grace de passer outre celuy terme y a grand esperance de parvenir en convalescence. Ie ne vous envoye point [vostre] revolution pource qu'il eut fallu trop long temps tant à la calculation que explication d'icelle. Quant à [vostre] fortune vous ne faillirez avoir d'enfans masles et femelles, et principalement masles que vous verrez en honneur et puissance grande, et auront de grandes charges. Quant au proces qu'avez à Paris de la detrousse des chevaux ils se defendent du privilege des guerres mais à la parfin vous en avrez bonne yssue. Quant à celuy contre les religieux d'Auberive ils seront du tout secourus, et favorisez et maintenus en leur premier estat. Et seroit bon qu'en cet endroit vous appointissiez car ils en auront du meilleur. De françoise de Pracontal il en y a de plus proches à la succession. Vray est qu'elle en amendera de beaucop. De [vostre] fortune. pour [vostre] avancement suivez les plus grands, et ne vous formalisez d'un costé ny d'autre. Tenez la voye de la messe plus tost qu'autrement. Quant à poison ne fault que craignez de cela. Quant à [vostre] maison de Saffre pour le seur y a un grandiss. thresor caché du temps de la guerre des Anglois. De vray le thresor qui y est est [sic] grand mais pour cette heure ie ne vous en puis rien resouldre pource que le loisir ne m'est donné de vous escrire au long: mais sera une autrefois Dieu aidant. Du proces que françoise de Pracontal a commencé a lencontre de son pere elle doibt laisser cela car son pere autant qu'il meure la recom-pensera amplement outre son mariage de beaucop. Quant à Bernarde de Cleron [vostre] soeur elle aura d'enfans masles et femelles et en brief. Quant au bruit quj a couru sur françoise de Pracontal contre son honneur le bruit est faux, meschant et deloyal. Et y a de femmes qui ont la langue plus venimeuse que aspie. Car pour le seur selon la figure astronomique grandiss. tort luy est fait par telles parolles semees. Et cela est autant faux et deloyal que celle qui a semé cela de la quelle ie ne veux dire le nom, qui a ainsi spargj telles parolles. Que sj elle faisoit son devoir elle luy viendroit demander pardon et mercy devant le monde. Par parolles de femmes, principalement de celles qui parlent plus qu'il n'est besoin, viennent grands maux. Une autre fois ie vous escrivay plus amplement Dieu aidant auquel ie prie.

Signeur de Saffre que vous doint santé vie longue et l'accomplissement entier de vos heroïques desirs. De Salon de Craux en Provence, ce XXV fevrier 1566.

Par [vostre] humble et obeissant
Michel Nostradamus conseilliers et medicin ordinayre du tres chrestien Roy de france [pres] vous hobeyr.

HISTORICAL NOTE

Nostradamus, who began his career at the court of Caterina de Medici in Paris, became famous for some successful prophecies, among them that of the assassination of King Henry IV in 1610. Between 1529 and 1533 Nostradamus studied medicine at the University of Montpellier.

This rare letter offers a valuable testimony to the credit and prestige enjoyed by horoscopes and astrology during the XVIth century. In it Nostradamus combined the horoscope of Monsieur de Saffre and his mother with medical advice, plain commonsense and the prediction of their futures. During the Renaissance astrological horoscopes were often regarded as an integral part of scientific endeavour. In connection with Nostradamus' controversial doctrine, Waller owned a copy of the rare pamphlet by Pierre Viret *Declaration des abus ignorances et seditions de Michel Nostradamus* ... (Avignon, 1558) (see *Bibliotheca Walleriana,* no. 20379).

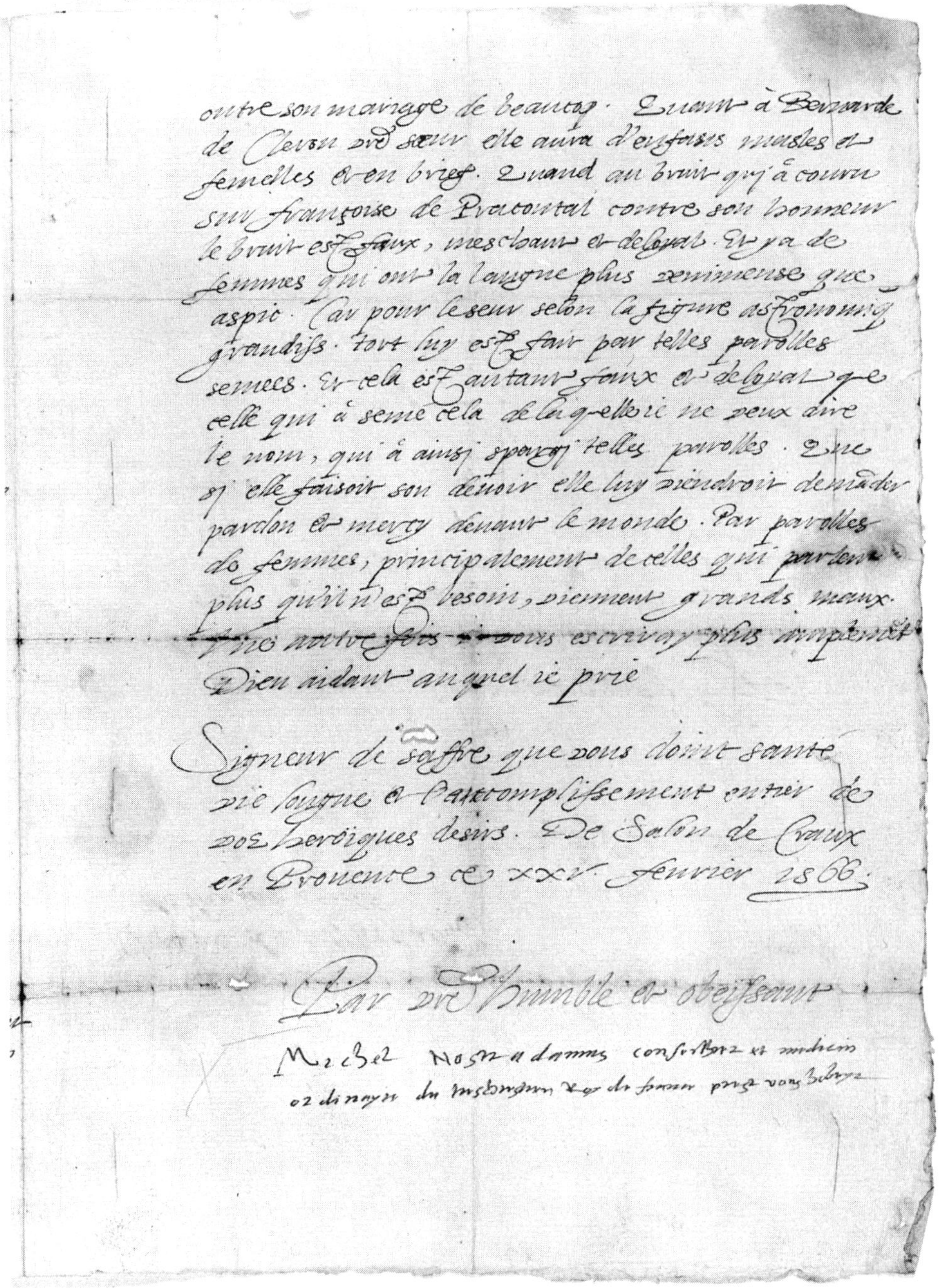

outre son mariage de beaucoup. Quant à Bernarde
de Cleron vre seur elle aura d'enfants masles et
femelles et en brief. Quand au bruit qui a couru
sur Françoise de Pracontal contre son honneur
le bruit est faux, meschant et deloyal. Et y a de
femmes qui ont la langue plus venimeuse que
aspic. Car pour le seur selon la figure astronomiq
grandiss. tout luy est fait par telles parolles
semees. Et cela est autant faux et deloyal que
celle qui a semé cela de laquelle ie ne veux dire
le nom, qui a ainsi spargi telles parolles. Que
si elle faisoit son devoir elle luy viendroit demander
pardon et mercy devant le monde. Par parolles
de femmes, principalement de celles qui parlent
plus qu'il n'est besoin, viennent grands maux.
Une autre fois ie vous escrivay plus amplement
Dieu aidant auquel ie prie

Signeur de Soffre que vous doint santé
vie longue et l'accomplissement entier de
vos heroiques desirs. De Salon de Craux
en Provence ce xxv. fevrier 1566.

Par vre humble et obeissant

Michel Nostradamus conseiller et medecin
ordinayre du [illegible]

Fig. 10.

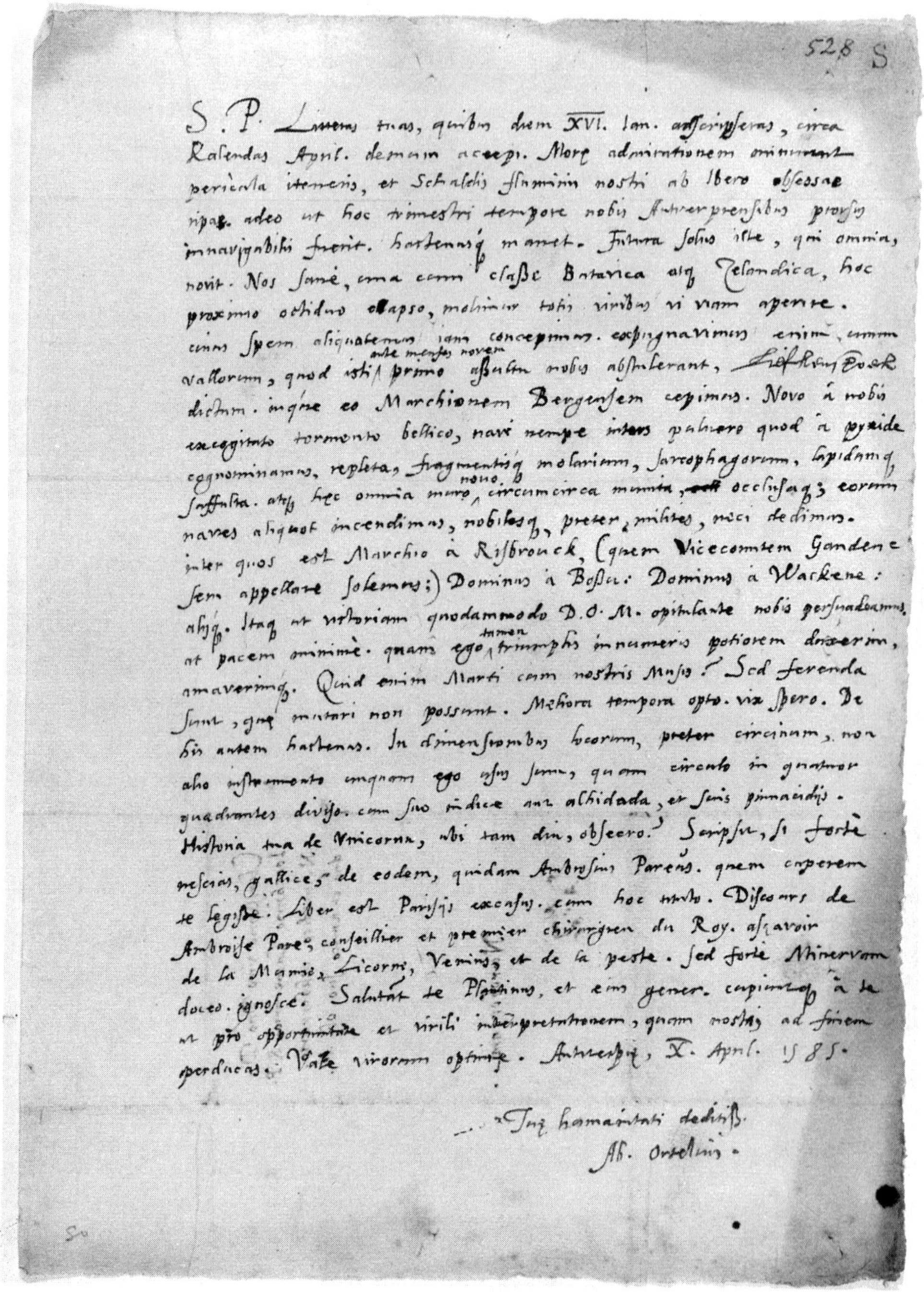

528 S

S. P. Literas tuas, quibus diem XVI. Ian. adscripseras, circa Kalendas April. demum accepi. Morae admirationem omittunt pericula itineris, et Schaldis fluminis nostri ab Ibero obsessae ripae. adeo ut hoc trimestri tempore nobis Antverpiensibus prorsus innavigabilis fuerit. hactenusque manet. Futura solus ille, qui omnia, novit. Nos sane, una cum classe Batavica atque Zelandica, hoc proximo octiduo elapso, molimur totis viribus vi viam aperire. cuius spem aliquatenus iam concepimus. expugnavimus enim, unum vallorum, quod isti ante menses novem primo assultu nobis abstulerant, Liefkens Hoek dictum. in quo eo Marchionem Bergensem cepimus. Novo a nobis excogitato tormento bellico, navi nempe interius pulvero quod a pyxide cognominamus, repleta, fragmentisque molarium, sarcophagorum, lapidumque suffulta. atque hec omnia muro novo, circumcirca munita, et occlusaque; eorum naves aliquot incendimus, nobilesque, preter milites, neci dedimus. inter quos est Marchio a Risbrouck, (quem Vicecomitem Gandensem appellare solemus;) Dominus a Bossu: Dominus a Warkene: aliique. Itaque ut victoriam quodammodo D.O.M. opitulante nobis persuadeamus, at pacem minime. quam ego tamen triumphis innumeris potiorem duxerim, amaverimque. Quid enim Marti cum nostris Musis? Sed ferenda sunt, quae mutari non possunt. Meliora tempora opto. vix spero. De his autem hactenus. In dimensionibus locorum, preter circinum, non alio instrumento umquam ego usus sum, quam circulo in quatuor quadrantes diviso. cum suo indice aut alhidada, et suis pinnacidiis. Historia tua de Unicornu, ubi tam diu, obsecro? Scripsit, si forte nescias, gallice, de eodem, quidam Andreas Pareus. quem cuperem te legisse. Liber est Parisiis excusus. cum hoc titulo. Discours de Ambroise Pare, conseillier et premier chirurgien du Roy. asçavoir de la Mumie, Licorne, Venins, et de la peste. Sed forte Minervam doceo. ignosce. Salutat te Plantinus, et eius gener. cupimusque a te ut pro opportunitate et virili interpretationem, quam nostras ad finem perducas. Vale virorum optime. Antverpiae, X. April. 1585.

Tuae humanitati deditiss.
Ab. Ortelius.

Fig. 11.

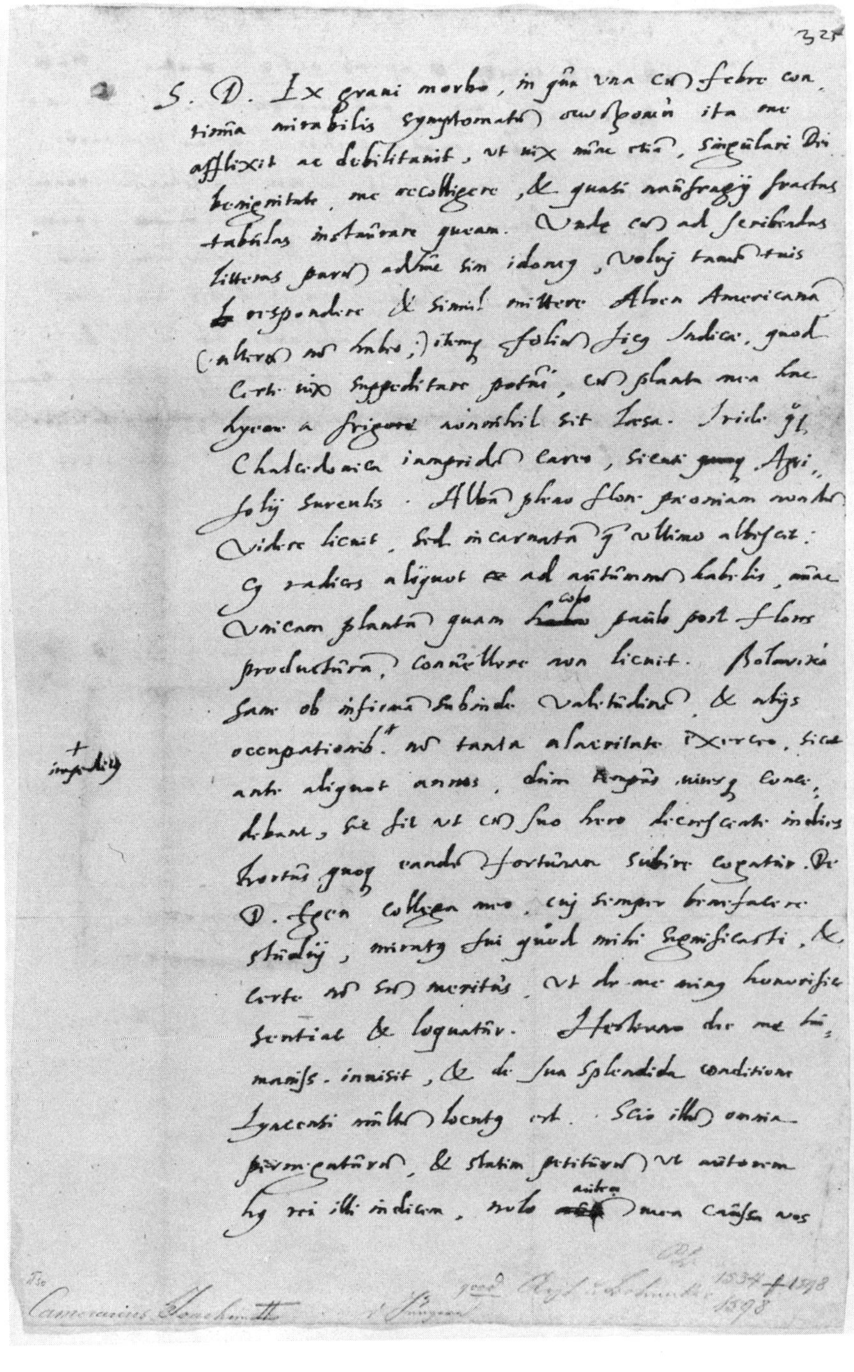

S. D. Ex gravi morbo, in quo una cum febre continua mirabilis symptomatum concursus ita me afflixit ac debilitavit, ut vix nunc etiam, singulari Dei benignitate, me recolligere, et quasi naufragii fractas tabulas instaurare queam. Unde cum ad scribendas litteras parum adhuc sim idoneus, volui tamen tuis respondere et simul mittere Aloen Americanam (alteram non habeo,) itemque folium Ficus Indicae, quod certe vix suppeditare potui, cum planta mea hac hyeme a frigore nonnihil sit laesa. Iridis [illegible] Chalcedonica iampridem careo, sicuti quoque Apii folii surculis. Album [illegible] florem [illegible] non videre licuit, sed incarnatum quod ultimo albescat. [illegible] radices aliquot ad autumnum habilis, unicam plantam quam cito paulo post flores producturam, communicare non licuit. Botanicen iam ob infirmam subinde valetudinem, et aliis occupationibus + non tanta alacritate exerceo, sicut ante aliquot annos, dum tempus minusque [illegible] debant, sic fit ut cum suo hero licentiati [illegible] hortus quoque eandem fortunam subire cogatur. De D. Egen collega meo, cuj semper [illegible] studij, miratusque sum quod mihi significasti, et certe non suis meritis, ut de me magis honorifice sentiat et loquatur. Herbarium [illegible] manuscr. invisit, et de sua splendida conditione Lyncenti multum locutus est. Scio illum omnia pervagaturum, et statim petiturum ut autorem huj rei illi indicem, nolo autem mea causa vos

+ impeditis

Fig. 12.

11.

ORTELIUS, ABRAHAM (1527-1598)

LETTER	Latin. Unpublished. The file contains an engraved frontispiece of Costanzo Landi's, *Complani Comitis, in Veterum Numismatum Romanorum Miscellanea Explicationes,* (Leiden, 1560), with Ortelius' inscription.
PLACE & DATE	Antwerp, 10.4.1585.
ADDRESSEE	Joachim Camerarius.
DIMENSIONS	29 x 19.2 cms.
FOLIATION	1 leaf.
SHELFMARK	Waller 649 F:2.
CONTENTS	

Ortelius, who in 1570 published the atlas *Theatrum orbis terrarum,* was together with Mercator one of the founders of modern cartography. His learning, however, in common with the scholars of the Renaissance was encyclopedic in scope. In this scientific letter Ortelius informs Camerarius about his work and asks for information about the latter's forthcoming book *Historia unicornum*. In this connection Ortelius informs Camerarius that the French surgeon Ambroise Paré has published an essay on the *Licorne* in his *Discours*. Ortelius also sends him the best regards of Christopher Plantin (the most famous printer of Antwerp during the Renaissance), thus showing the intimate relation between the world of learning and the printers during the XVIIth century.

12.

CAMERARIUS, JOACHIM (1534-1598)

LETTER	Latin. Unpublished. The same file contains two more letters dated 1577 and 1590.
PLACE & DATE	Nürnberg, 16.4.1598.
DIMENSIONS	31.2 x 19.1 cms.
FOLIATION	1 leaf.
SHELFMARK	Waller 651 D:2.
CONTENTS	

Letter dealing with medical and botanical matters. Camerarius is discussing the therapeutical qualities of some exotic plants hitherto unknown in Europe.

HISTORICAL NOTE

Joachim Camerarius studied medicine at the University of Bologna during the 1560's and corresponded with the most prominent physicians of the period. While in Nürnberg, he successfully applied his knowledge of the vegetable kingdom to the preparation of medical remedies.

13.

RAVE [RAVIUS], JOHANN (1578-1621) AND SANTORIO SANTORIO (1561-1636)

DOCTORAL DIPLOMA
IN MEDICINE — Latin.
PLACE & DATE — Padua, 1618.
DIMENSIONS — 23.2 x 16.7 cms.
FOLIATION — 10 leaves on vellum.
ILLUMINATED
BINDING — Decorated contemporary brown morocco.
SHELFMARK — Waller Red Morocco Box.
HISTORICAL NOTE

Friend of Galileo and prominent physician, Santorio Santorio taught medicine at Padua University between 1615 and 1624. This beautifully illuminated Diploma bears Santorio's signature and the coat of arms of J. Rave, a German physician.

Among wealthy students it was common to adorn their doctoral diplomas with expensive binding and illumination. The second illuminated leaf of the Diploma (see Fig. 13) shows on the lower part a dissection of a human body and on the sides two scientists holding medical and astronomical instruments.

14.

BRAHE, TYCHO (1546-1601)

ALBUM AMICORUM
(AUTOGRAPH BOOK) — 1 leaf. Unpublished?
PLACE & DATE — Dresden, 10.11.1598.
DIMENSIONS — 18.4 x 13.2 cms.

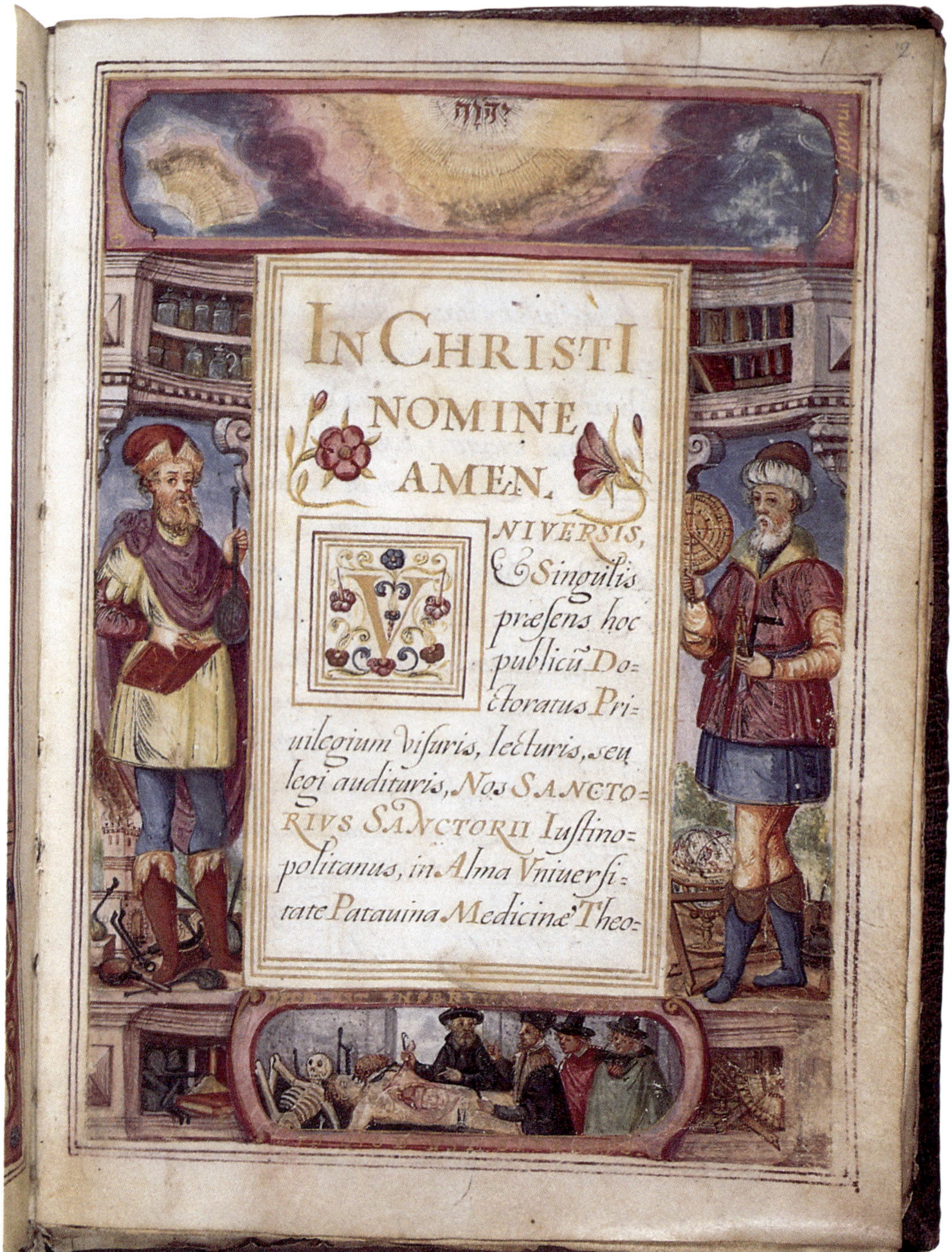
IN CHRISTI
NOMINE
AMEN.
VNIVERSIS, & Singulis præsens hoc publicũ Doctoratus Priuilegium visuris, lecturis, seu legi audituris, Nos SANCTORIVS SANCTORII Iustinopolitanus, in Alma Vniuersitate Patauina Medicinæ Theo-

Fig. 13.

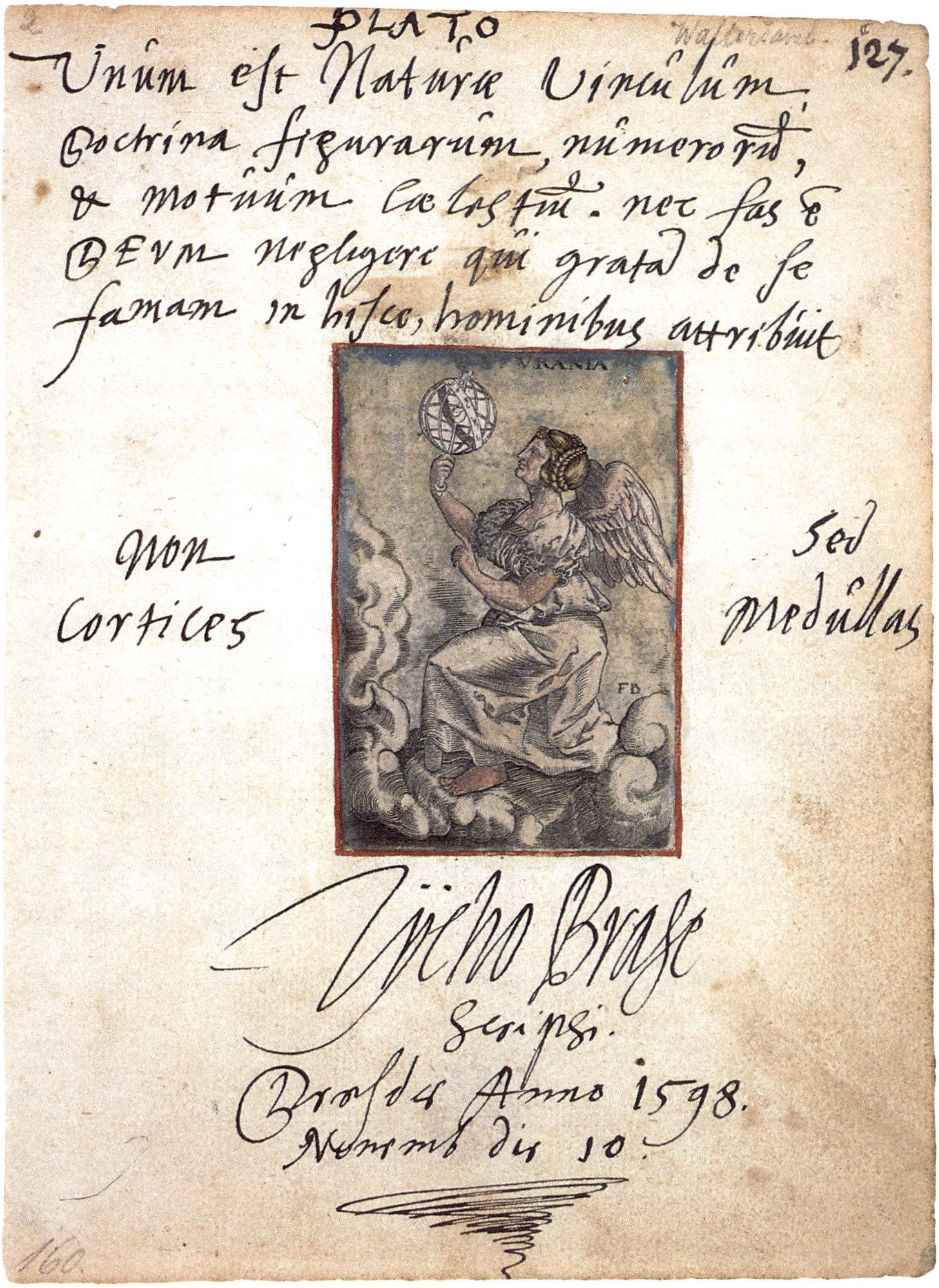

PLATO
Unum est Naturae Vinculum,
Doctrina figurarum, numerorum,
& motuum caelestium. nec fas est
DEUM negligere qui gratam de se
famam in hisce, hominibus attribuit

Non cortices sed medullas

Tycho Brahe
scripsi.
Dresdae Anno 1598.
Novemb. die 10.

Fig. 14.

ILLUSTRATED

SHELFMARK Waller Delikatesskaps 2.

HISTORICAL NOTE During the Renaissance it was common among the learned, especially in Germany and in the Nordic countries, to invite distinguished guests to sign the "Album amicorum" or "Philotheca". This tradition belonged both to the universities and to the private learned world.

This interesting leaf with the signature of the famous astronomer Tycho Brahe gives Brahe's free translation of a passage from Plato (dialogue *Epinomis* 991D-992A):

> "Unum est Naturae Vinculum. Doctrina
> figurarum, numerorum, et motuum caelestium.
> nec fas est Deum negligere qui gratam de se
> famam in hisce, hominibus attribuit."

The leaf also bear the motto:

> "non cortices sed medullas".

At the center an engraving of Urania (the Muse who presided over astronomy), possibly the emblem of Brahe's astronomical observatory in Uraniaborg.

15.

KEPLER, JOHANN (1571-1630)

ALBUM AMICORUM

(AUTOGRAPH BOOK) 1 leaf. Published in Johannes Kepler, *Gesammelte Werke*, vol. XIX, (München, 1975), p. 384. The file also contains a letter dated 4.11.1626.

PLACE & DATE Ulm, 11-21.11.1627.

DIMENSIONS 14.8 x 9.1 cms.

SHELFMARK Waller Delikatesskaps 2.

TRANSCRIPTION

Sub juga non potuit Terram unam mittere planè:
O curas hominum, ô quantum est in rebus inane;

Joannes Keplerus, Imp: Caes: Ferdinandj II.
Ordinumque Austriae Supr-Anisanae Mathe-
maticus scripsi Vlmae Suevicae, oculis et ani-
mo peregrinatus in regionibus Solaribus et
contemplatione nubium Solarium per Tubum
opticum $\frac{11}{21}$ Novembris anno MDCXXVII.

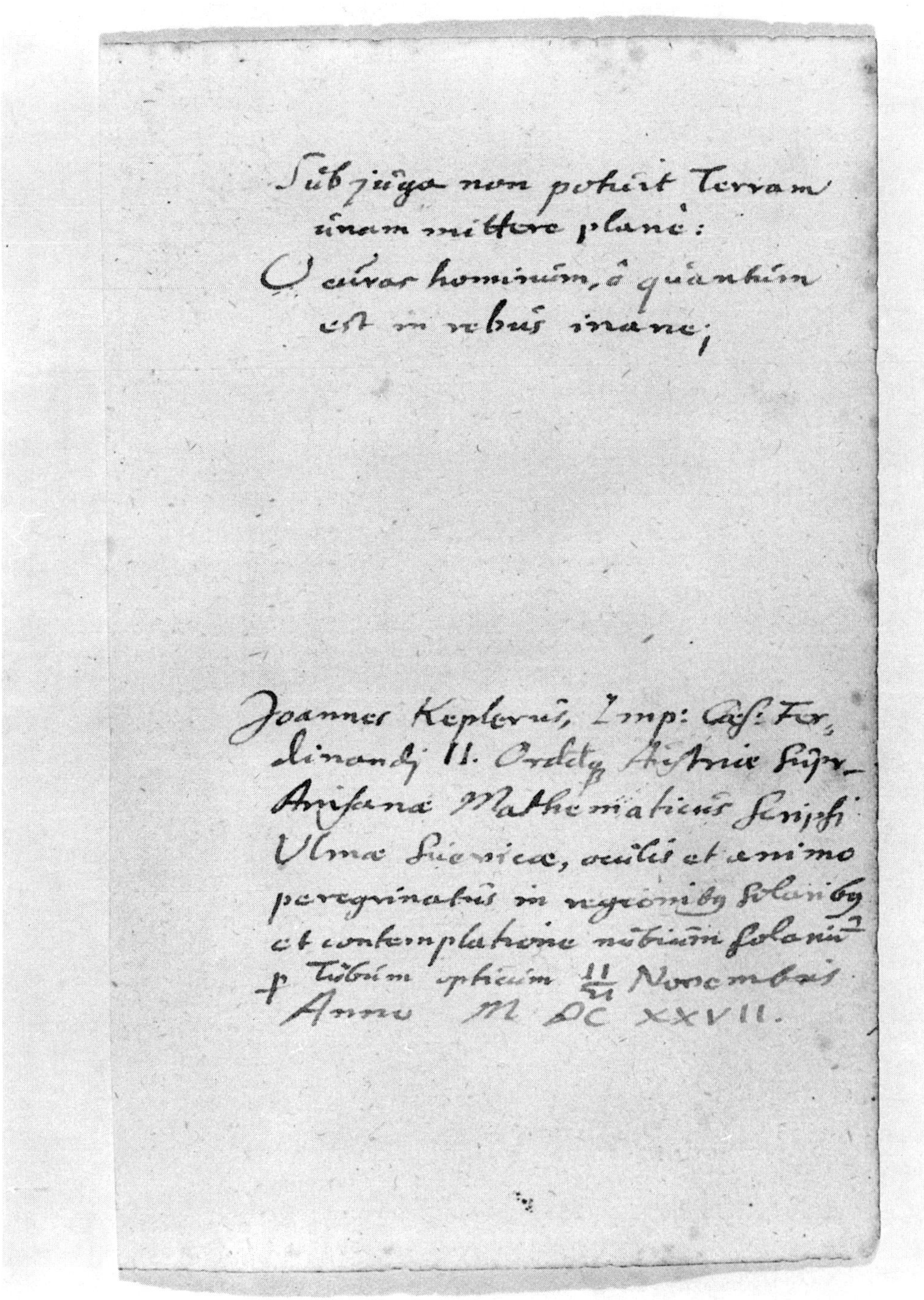
Sub juga non potuit terram
unam mittere plane:
O curas hominum, ô quantum
est in rebus inane;

Joannes Keplerus, Imp: Caes: Fer-
dinandi II. Ordd. Austriae Supr-
Anisanae Mathematicus scripsi
Ulmae Suevicae, oculis et animo
peregrinatus in regionibus solaribus
et contemplatione nubium solarium
p Tubum opticum 11/21 Novembris
Anno M DC XXVII.

Fig. 15.

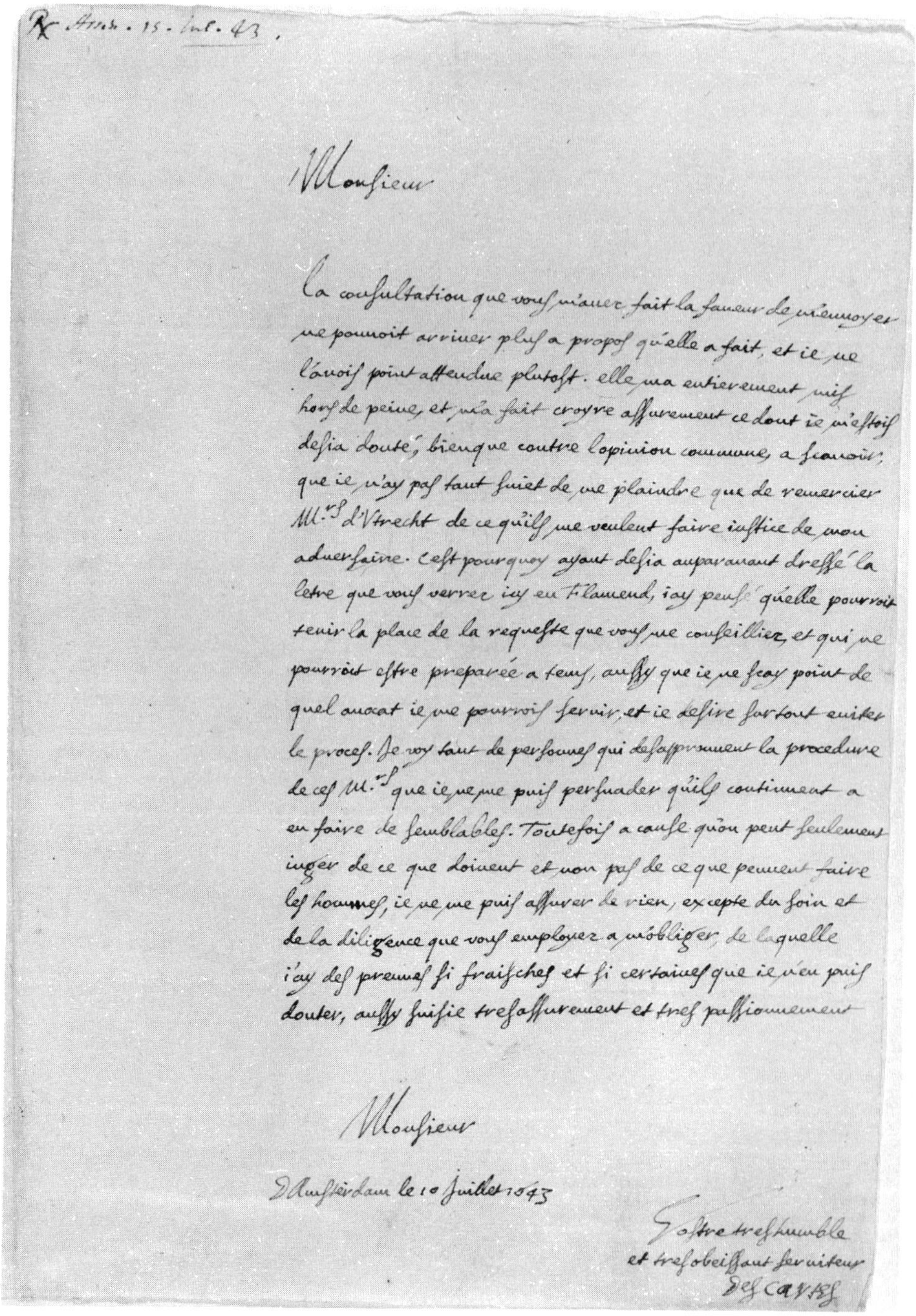

Monsieur

La consultation que vous m'avez fait la faveur de m'envoyer ne pouvoit arriver plus a propos qu'elle a fait, et ie ne l'avois point attendue plutost. elle m'a entierement mis hors de peine, et m'a fait croyre assurement ce dont ie m'estois desia douté, bienque contre l'opinion commune, a scavoir, que ie n'ay pas tant suiet de me plaindre que de remercier Mrs d'Vtrecht de ce qu'ils me veulent faire iustice de mon adversaire. C'est pourquoy ayant desia auparavant dressé la letre que vous verrez icy en Flamend, i'ay pensé qu'elle pourroit tenir la place de la requeste que vous me conseilliez, et qui ne pourroit estre preparée a tems, aussy que ie ne scay point de quel avocat ie me pourrois servir, et ie desire surtout eviter le proces. Je voy tant de personnes qui desapprouvent la procedure de ces Mrs que ie ne me puis persuader qu'ils continuent a en faire de semblables. Toutefois a cause qu'on peut seulement iuger de ce que doivent et non pas de ce que peuvent faire les hommes, ie ne me puis assurer de rien, excepte du soin et de la diligence que vous employez a m'obliger, de laquelle i'ay des preuves si fraisches et si certaines que ie n'en puis douter, aussy suisie tresassurement et tres passionement

Monsieur

D'Amsterdam le 10 Juillet 1643

Vostre treshumble
et tresobeissant serviteur
DesCartes

Fig. 16.

Fig. 17.

16.

DESCARTES, RENÉ (1596-1650)

LETTER	French. Published.
PLACE & DATE	10.7.1643.
ADDRESSEE	Const. Huyghens.
DIMENSIONS	30.1 x 20.7 cms.
FOLIATION	2 leaves.
SHELFMARK	Waller Delikatesskaps 2.
HISTORICAL NOTE	

This letter disappeared when the collection it belonged to was transferred to the University Library of Leiden in 1862. Fortunately it was published prior to that date and Charles Adam and Paul Tannery reprinted its text in the fourth volume of the *Oeuvres de Descartes,* (Paris, 1901), pp. 13-14. The original differs from the published version only in the punctuation.

17.

[DESCARTES, RENÉ (1596-1650)] AND BOYLE, ROBERT (1627-1691)

ENGRAVED	
PORTRAIT	Interesting engraving depicting Descartes and signed by the British chemist, Robert Boyle. Two more autograph letters by Boyle are kept in Waller 648 A:1.
DATE	XVIIth century.
DIMENSIONS	18.5 x 15.5 cms.
LOCATION	Waller Collection of engravings: file "Boyle".
HISTORICAL NOTE	

During the XVIIth and XVIIIth centuries the exchange and collecting of portraits of scientists, philosophers and the learned was extremely common. To exchange and collect portraits was intended to enlarge the sources of mutual knowledge among the learned. This rare and little known engraving shows Descartes writing a work while receiving light from heaven. The room is filled with scientific instruments and the tree behind the head of Descartes probably symbolises the tree of knowledge. Naturally it is the autograph of Robert Boyle, who supported Descartes' mechanical philosophy, that gives this portrait great historical value.

18.

DU BOIS, JACQUES [JACOBUS, SYLVIUS] (1478-1555); FABRICIUS AB AQUAPENDENTE, HIERONYMUS (1533-1619); HARVEY, WILLIAM (1578-1657); MEAD, RICHARD (1673-1754)

BOOK — Medical work by Du Bois with marginal notes by Fabricius and the signatures by Harvey and Mead.

TITLE — *De febribus, commentarius ex libris aliquot Hippocratis Galeni, parte plurima selectus ... Cui adiunximus Camilli Thomai medici commentariolum. De febrium humoralium, hecticarum, & qua pestilentialum dignotione, ac curatione,* (Venetiis, Arrivabemus, 1555).

DIMENSIONS — 14.1 x 10 cms.

FOLIATION — (15) + 71 leaves.

SHELFMARK — Bibliotheca Walleriana Rar. 2582.

HISTORICAL NOTE

Disciple of Gabriele Falloppio, Fabricius became Professor of Surgery and Anatomy at the University of Padua in 1565. In his physiological and embryological works, Fabricius laid the basis for Harvey's investigation into the circulation of the blood. Harvey arrived in Padua in 1599 where he studied under the supervision of Fabricius and as early as 1602 returned to England. These biographical circumstances pose some questions concerning the history of this exemplar. The book was purchased or received by Harvey in 1621. Fabricius died two years before and unless it was written in his will that the book was to be given to Harvey, it is difficult to be sure about the original ownership. On the first leaf of the book there is the inscription "Will. Harvey 1621"; on the last leaf of the text those of "R. Mead" and "Geronimo Fabrizio". The signature supposedly by Fabricius does not seem to be the same as the handwriting of the marginal notes. According to Waller and later Sallander, "the first owner of this book was Harvey's teacher in Padua, Hieronymus ab Aquapendente, who on almost every page has annotated the text with marginalia". However, on a closer examination of the manuscript notes it is quite clear that the marginalia belong to two distinct hands one later than the other; one of the hands seems to be of the XVIth century and it might belong to Fabricius. At the end of the text there are two leaves of medical prescriptions possibly written by Fabricius. As for the other hand it can be stated with confidence that it does not belong to Harvey.

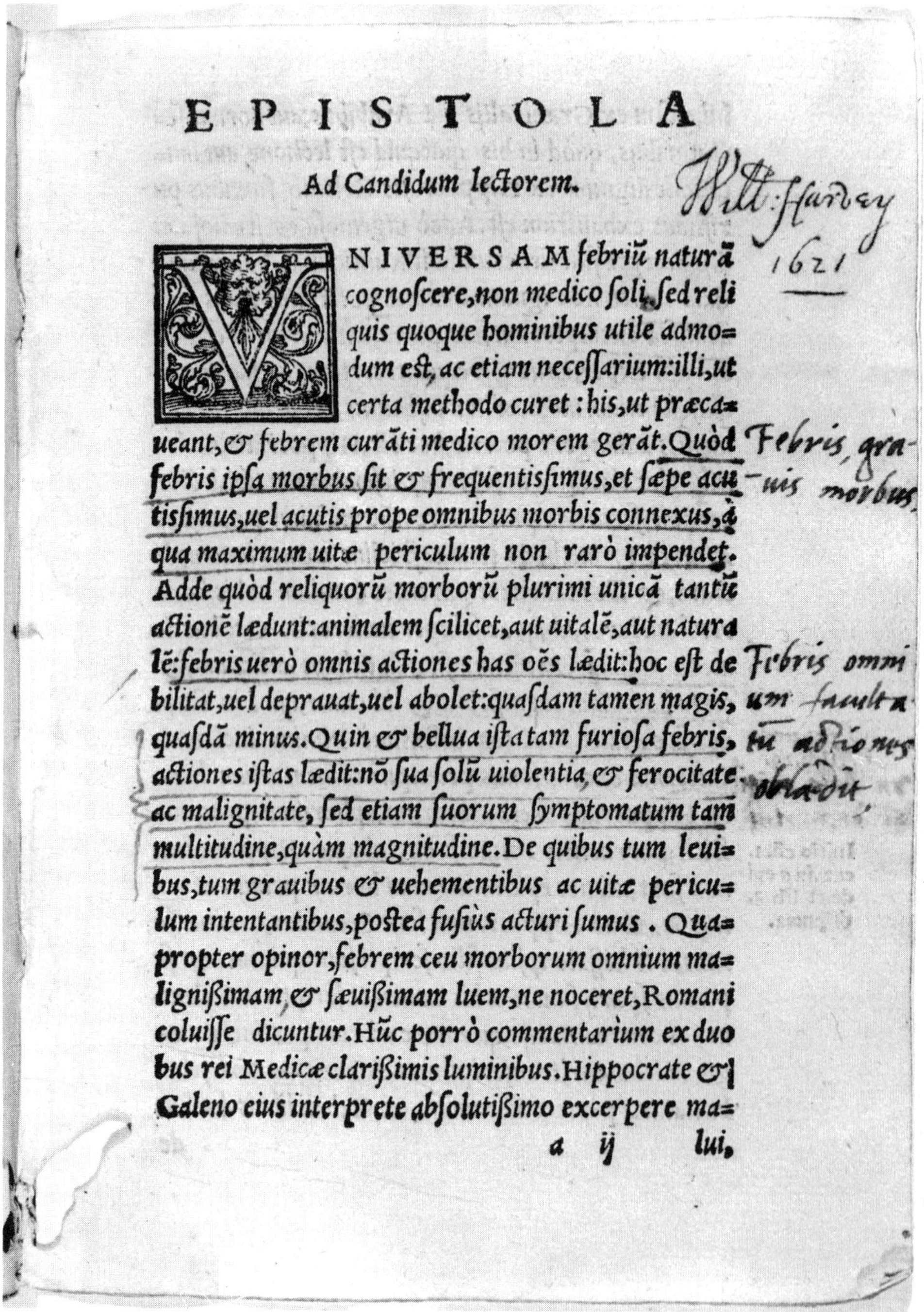

EPISTOLA

Ad Candidum lectorem.

VNIVERSAM febriū naturā cognoscere, non medico soli, sed reliquis quoque hominibus utile admodum est, ac etiam necessarium: illi, ut certa methodo curet: his, ut præcaueant, & febrem curāti medico morem gerāt. Quòd febris ipsa morbus sit & frequentissimus, et sæpe acutissimus, uel acutis prope omnibus morbis connexus, à qua maximum uitæ periculum non rarò impendet. Adde quòd reliquorū morborū plurimi unicā tantū actionē lædunt: animalem scilicet, aut uitalē, aut naturalē: febris uerò omnis actiones has oēs lædit: hoc est debilitat, uel deprauat, uel abolet: quasdam tamen magis, quasdā minus. Quin & bellua ista tam furiosa febris, actiones istas lædit: nō sua solū uiolentia, & ferocitate ac malignitate, sed etiam suorum symptomatum tam multitudine, quàm magnitudine. De quibus tum leuibus, tum grauibus & uehementibus ac uitæ periculum intentantibus, postea fusiùs acturi sumus. Quapropter opinor, febrem ceu morborum omnium malignissimam, & sæuissimam luem, ne noceret, Romani coluisse dicuntur. Hūc porrò commentarium ex duobus rei Medicæ clarissimis luminibus, Hippocrate & Galeno eius interprete absolutissimo excerpere ma-

a ij lui,

Fig. 18 a.

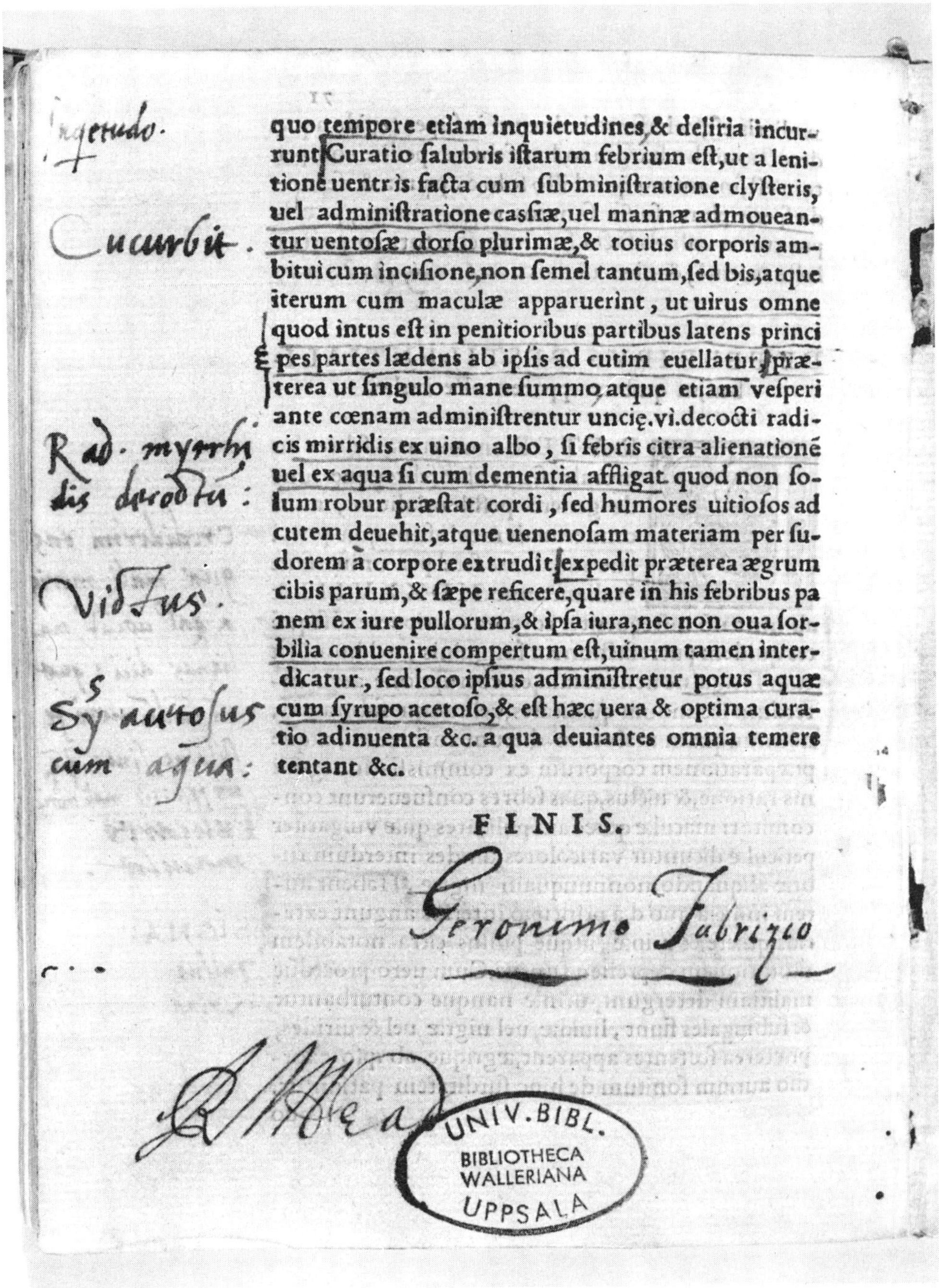
quo tempore etiam inquietudines,& deliria incurrunt. Curatio ſalubris iſtarum febrium eſt,ut a lenitione uentris facta cum ſubminiſtratione clyſteris, uel adminiſtratione caſſiæ,uel mannæ admoueantur uentoſæ dorſo plurimæ,& totius corporis ambitui cum inciſione,non ſemel tantum,ſed bis,atque iterum cum maculæ apparuerint, ut uirus omne quod intus eſt in penitioribus partibus latens principes partes lædens ab ipſis ad cutim euellatur, præterea ut ſingulo mane ſummo,atque etiam veſperi ante cœnam adminiſtrentur unciæ.vi.decocti radicis mirridis ex uino albo, ſi febris citra alienationẽ uel ex aqua ſi cum dementia affligat. quod non ſolum robur præſtat cordi, ſed humores uitioſos ad cutem deuehit,atque uenenoſam materiam per ſudorem à corpore extrudit. expedit præterea ægrum cibis parum,& ſæpe reficere,quare in his febribus panem ex iure pullorum,& ipſa iura,nec non oua ſorbilia conuenire compertum eſt,uinum tamen interdicatur, ſed loco ipſius adminiſtretur potus aquæ cum ſyrupo acetoſo,& eſt hæc uera & optima curatio adinuenta &c. a qua deuiantes omnia temere tentant &c.

FINIS.

Fig. 18 b.

19.

[GALILEI, GALILEO (1564-1642) AND KEPLER, JOHANN (1571-1630)]

LETTER Italian and Latin. Published. A description of the contents of the whole file "Galileo" is given at the end of the historical note of item No. 22.

PLACE & DATE Rome, 22.4.1611.

ADDRESSEE [Filippo Salviati in Florence.]

DIMENSIONS 28.6 x 20.8 cms.

FOLIATION 2 leaves.

SHELFMARK Waller 650 A:2.

HISTORICAL NOTE

In 1841 Pietro Bigazzi published this letter in his *Due lettere di Galileo Galilei ed una del Keplero inedite,* (Florence). Since he did not have access to the original, Favaro reprinted the version given by Bigazzi in the eleventh volume of *Le opere di Galileo Galilei,* (Florence, 1901), pp. 89-91. The letter includes Galileo's transcription of a letter by the astronomer Johann Kepler.

20.

[GALILEI, GALILEO (1564-1642)]

LETTER Italian. Published.

DATE [1615.]

ADDRESSEE Cristina di Lorena.

DIMENSIONS 27.3 x 19.6 cms.

FOLIATION 4 leaves.

SHELFMARK Waller 650 A:2.

CONTENTS

Fragment of a copy of the letter Galileo wrote to Cristina di Lorena (the whole letter was published by Favaro in *Le opere di Galileo Galilei,* vol. 5, (Florence, 1895), pp. 309-348). With the exception of a few changes in the punctuation, this version is identical to that published by Favaro. However, at the end of the manuscript is added a latin quotation from Plato which was missing in the version used by Favaro. The full quotation is:

> "Naturam rerum invenire difficile:
> et ubi inveneris, indicare in
> vulgus, nefas. Plato"

The fragment of the letter begins with the following words: "Resta finalmente che consideriamo quanto sia vero, che il luogo di Giosuè si possa prendere senza alterare il puro significato delle parole, e come possa essere, che obbedendo il sole al Comandamento di Giosuè, che fu che egli si fermasse, ne potesse da ciò seguire che il giorno per molto si prolungasse." The text covers the concluding part of the letter *(Le opere,* cit., pp. 343-348).

21.

BOLOGNETTI, GIORGIO (VESCOVO D'ASCOLI)

LETTER	Italian. Published in *Le opere di Galileo Galilei,* vol. 15, (Florence, 1904), p. 242.
PLACE & DATE	Florence, 27.8.1633.
ADDRESSEE	Pietro Niccolini.
DIMENSIONS	27.1 x 19.3 cms.
FOLIATION	2 leaves.
SHELFMARK	Waller 650 A:2.
CONTENTS	Letter in which Giorgio Bolognini, who was Bishop of Ascoli from 1630, encloses a copy of Galileo's abjuration and a copy of the sentence of the Catholic Church against Galileo's heretical views to Pietro Niccolini (who was Bishop of Florence).

22.

[GALILEI, GALILEO, (1564-1642)]

MANUSCRIPT	Italian. Unbound. Probably this is the copy of the sentence against Galileo and of his abjuration enclosed by Bolognetti (see item No. 21) to Niccolini. Another copy was published in *Le Opere di Galileo Galilei,* vol. 19, (Florence, 1907), pp. 403-407.
DATE	1633.
DIMENSIONS	27.1 x 19.2 cms.
FOLIATION	6 leaves.
SHELFMARK	Waller 650 A:2.
HISTORICAL NOTE	

In August 1632 the sale of Galileo's *Dialogo* was prohibited and its contents were examined by a special commission who reported as follows:

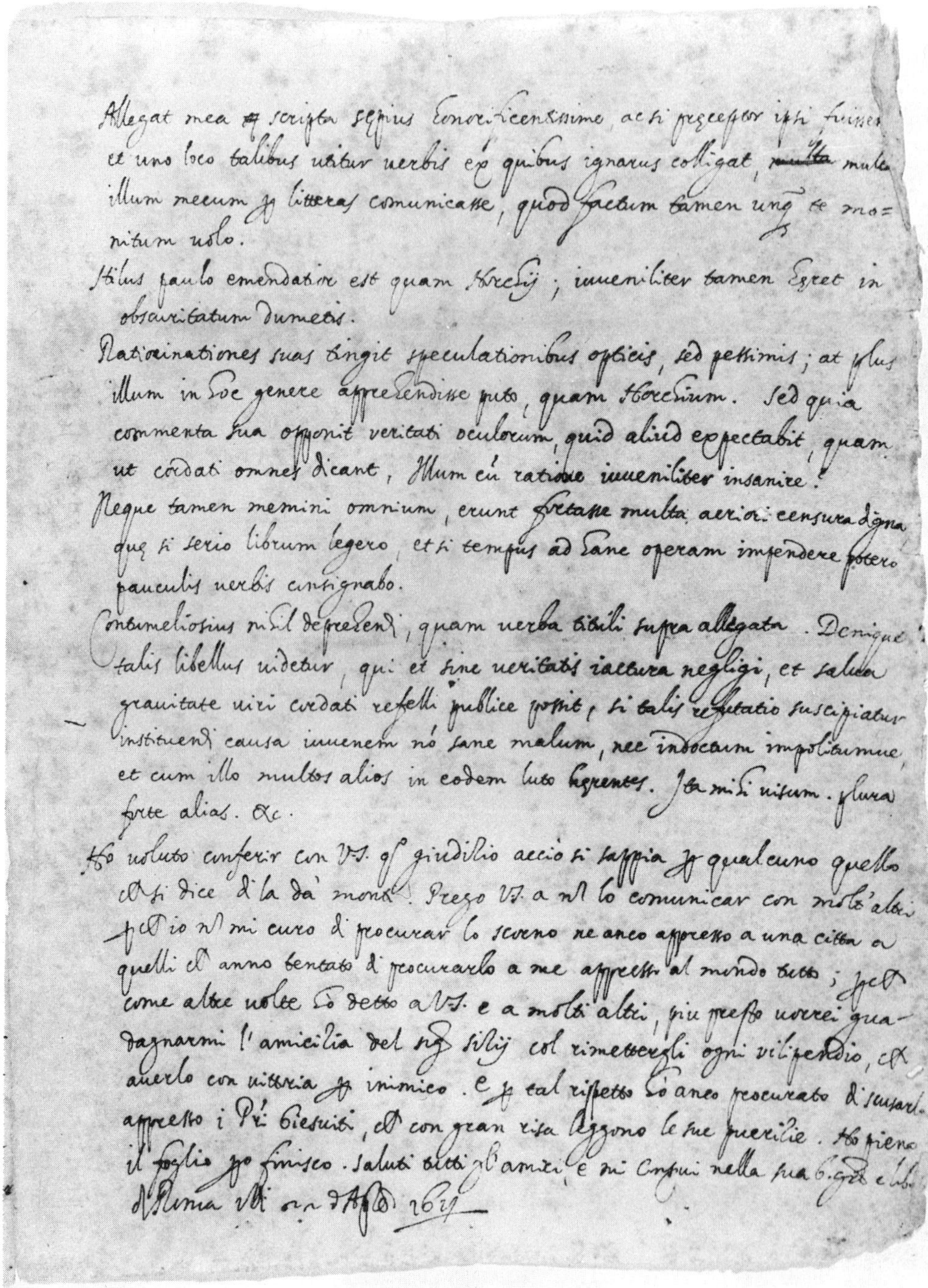
Allegat mea scripta sepius honorificentissime, ac si preceptor ipsi fuisset
et uno loco talibus utitur verbis ex quibus ignarus colligat, ~~multa~~ multa
illum mecum per litteras comunicasse, quod factum tamen unq̄ te mo=
nitum volo.
Stilus paulo emendatior est quam Horcij; iuveniliter tamen hæret in
obscuritatum dumetis.
Ratiocinationes suas tingit speculationibus opticis, sed pessimis; at plus
illum in hoc genere apprehendisse puto, quam Horcium. Sed quia
commenta sua opponit veritati oculorum, quid aliud expectabit, quam
ut cordati omnes dicant, illum cū ratione iuveniliter insanire?
Neque tamen memini omnium, erunt forsitan multa acriori censura digna
quę si serio librum legero, et si tempus ad hanc operam impendere potero
pauculis verbis consignabo.
Contumeliosius nihil deprehendi, quam verba tibili supra allegata. Denique
talis libellus videtur, qui et sine veritatis iactura negligi, et salva
gravitate viri cordati refelli publice possit, si talis refutatio suscipiatur
instituendi causa iuvenem nō sane malum, nec indoctum impolitumue,
et cum illo multos alios in eodem luto hærentes. Ita mihi visum. plura
forte alias. &c.
Ho voluto conferir con V.S. q(ue)sto giuditio accio si sappia per qualcuno quello
che si dice di la dai monti. Prego V.S. a non lo comunicar con molti altri
perche io non mi curo di procurar lo scorno ne anco appresso a una citta a
quelli che anno tentato di procurarlo a me appresso al mondo tutto; perche
come altre volte ho detto a V.S. e a molti altri, piu presto vorrei gua-
dagnarmi l'amicitia del sig.r Sitij col rimettergli ogni vilipendio, che
averlo con vittoria per inimico. E per tal rispetto ho anco procurato di scusarlo
appresso i P.ri Giesuiti, che con gran risa leggono le sue puerilie. Ho pieno
il foglio pero finisco. saluti tutti gl'amici e mi conservi nella sua b. gr(ati)a e ...
di Roma l'ult.o d'Aple 1611

Fig. 19.

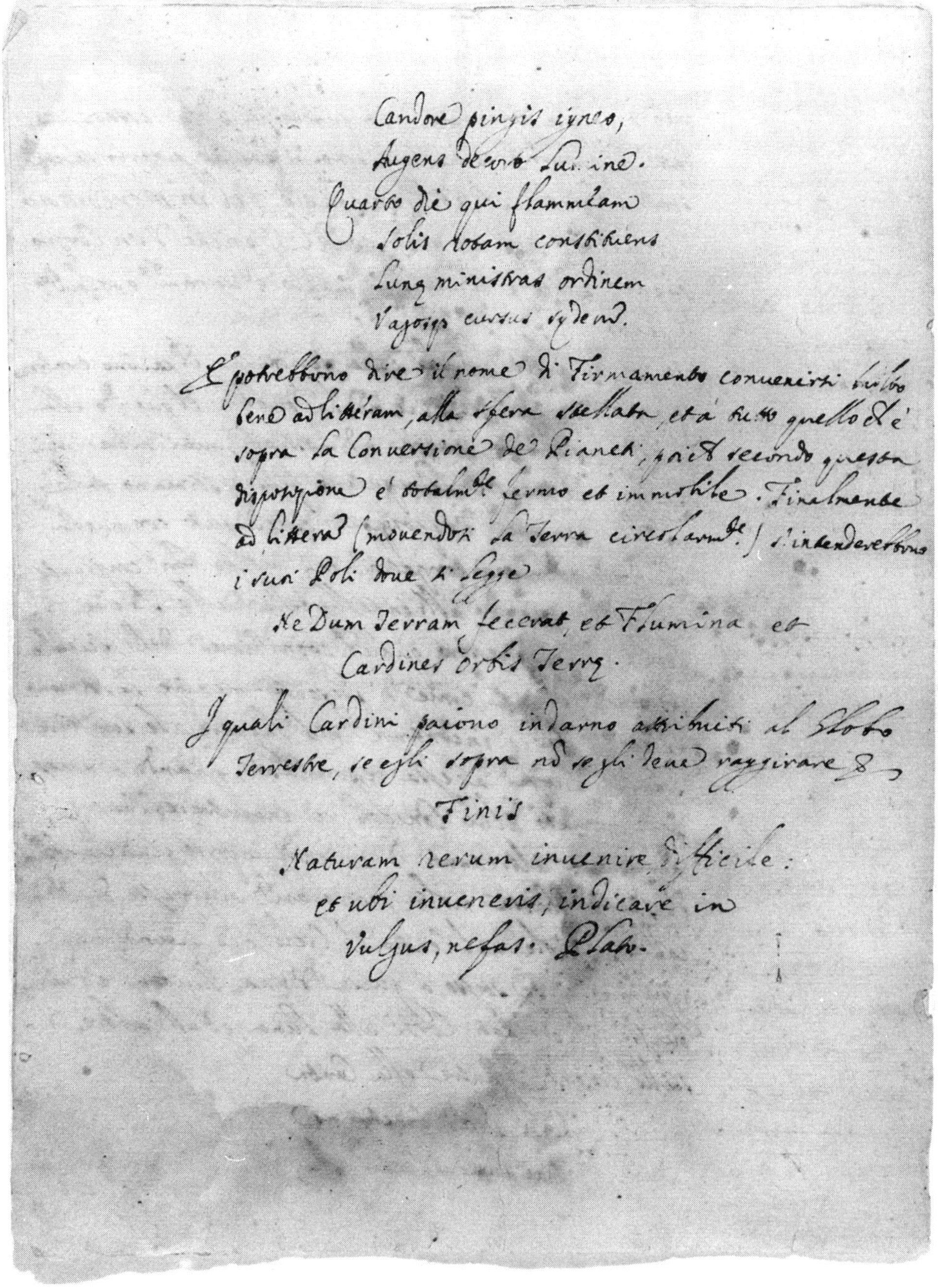

Candore pingis igneo,
Augens decoro lumine.
Quarto die qui flammeam
Solis rotam constituens
Lunę ministras ordinem
Vagosque cursus sydeum.

E potrebbono dire il nome di Firmamento convenirsi tutto bene ad litteram, alla sfera stellata, et à tutto quello che è sopra la Conversione de' Pianeti; poichè secondo questa dispositione è totalm.te fermo et immobile. Finalmente ad litteram (movendosi la Terra circolarm.te) s'intenderebbono i suoi Poli dove si legge

Ne Dum Terram fecerat, et Flumina et
Cardines Orbis Terrę.

I quali Cardini paiono indarno attribuiti al Globo Terrestre, se egli sopra essi non se gli deve raggirare &c.

Finis

Naturam rerum invenire difficile:
et ubi inveneris, indicare in
vulgus, nefas. Plato.

Fig. 20.

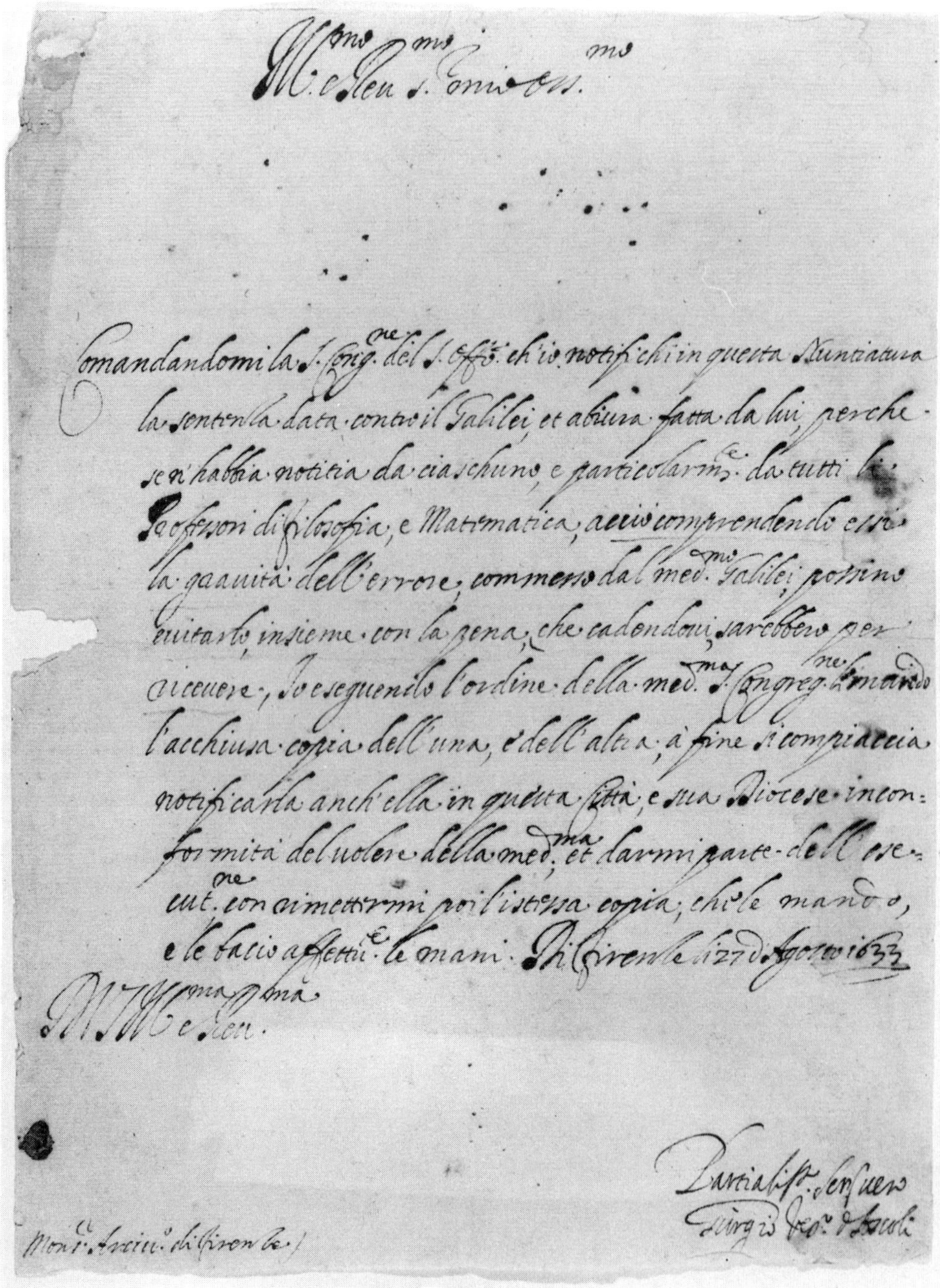

Ill.mo e Rev.mo S.r mio Oss.mo

Comandandomi la S. Congreg.ne del S. Off.o ch'io notifichi in questa Nuntiatura la sentenza data contro il Galilei et abiura fatta da lui, perche se n'habbia notitia da ciascheduno, e particolarm.te da tutti li Professori di Filosofia, e Matematica, accio comprendendo essi la gravità dell'errore, commesso dal med.mo Galilei, possino evitarlo insieme con la pena, che cadendovi sarebbero per riceuere, Io eseguendo l'ordine della med.ma S. Congreg.ne le mando l'acchiusa copia dell'una, e dell'altra, a fine si compiaccia notificarla anch'ella in questa Città, e sua Diocese in conformità del volere della med.ma, et darmi parte dell'esecut.ne con rimettermi poi l'istessa copia, ch'io le mando, e le bacio affettuos.te le mani. Di Firenze li 27 di Agosto 1633

Di V.S. Ill.ma e Rev.ma

Partialiss.mo Ser.re vero
Giorgio Vesc.o d'Ascoli

Mons.r Arciv.o di Firenze

Fig. 21.

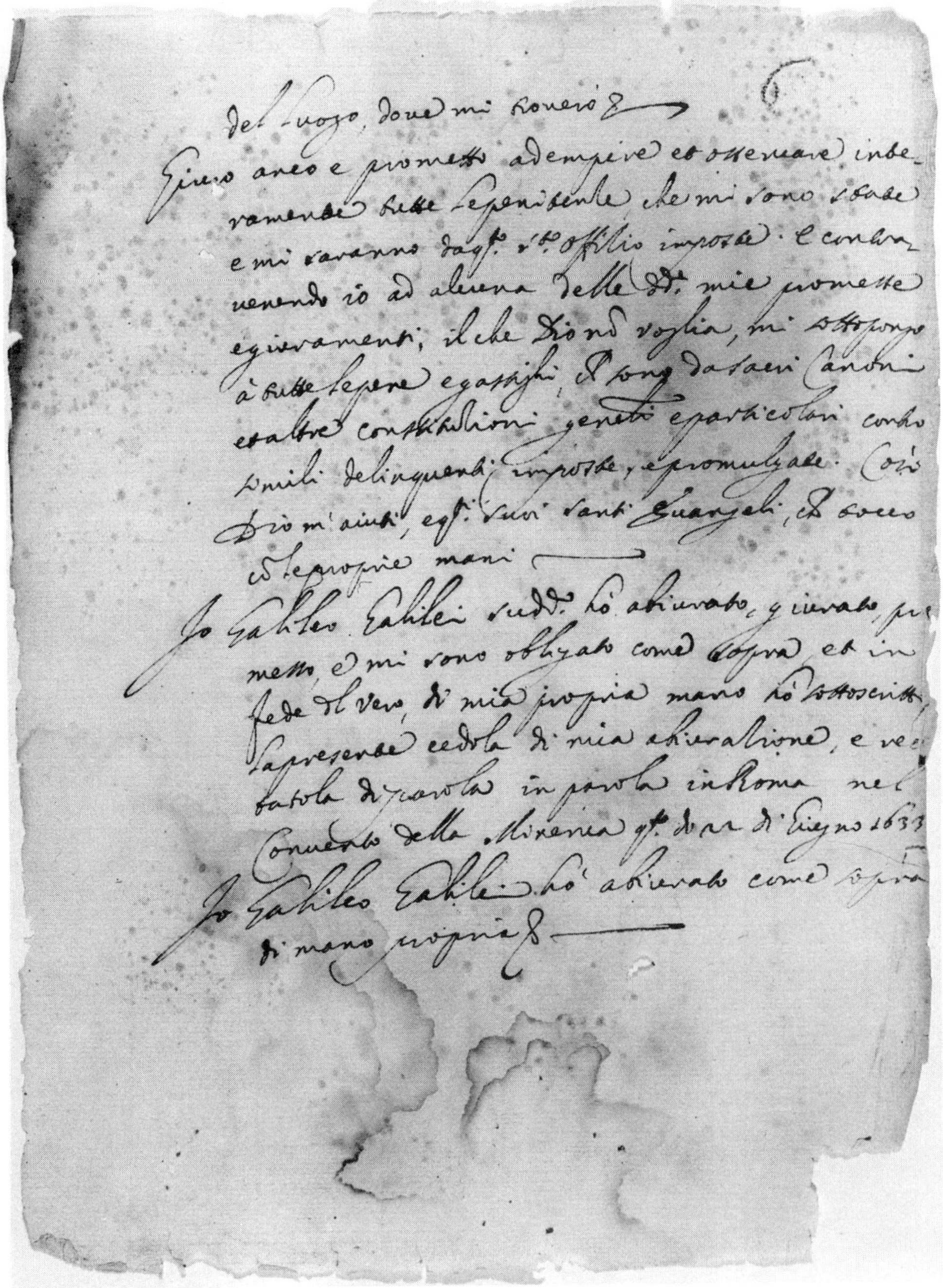

6

del luogo, doue mi trouerò
Giuro anco e prometto adempire et osseruare inte-
ramente tutte le penitenze che mi sono state
e mi saranno da q.sto S.to Offitio imposte. E contra-
uenendo io ad alcuna delle dd.e mie promesse
e giuramenti, il che Dio nõ voglia, mi sottopongo
à tutte le pene e castighi, che sono da sacri Canoni
et altre constitutioni generali e particolari contro
simili delinquenti imposte, e promulgate. Così
Dio m'aiuti, e q.sti suoi santi Euangeli, che tocco
con le proprie mani
Io Galileo Galilei sudd.o hò abiurato, giurato, pro-
messo, e mi sono obligato come sopra, et in
fede del vero, di mia propria mano hò sottoscritta
la presente cedola di mia abiuratione, e re-
citata di parola in parola in Roma nel
Conuento della Minerua q.sto dì 22 di Giugno 1633
Io Galileo Galilei hò abiurato come sopra
di mano propria

Fig. 22.

"Galileo has trangressed orders in deviating from the hypothetical standpoint, by maintaining *decidedly* that the earth moves and the sun is stationary."
"He has erroneously ascribed the phenomena of the tides to the stability of the sun and the motion of the earth, which is not true."
"He has been deceitfully silent about the command laid upon him to relinquish altogether the opinion that the sun is the centre of the world and immovable and that the earth moves, nor henceforth to hold, teach, or defend it in any way whatsoever, verbally or in writing."

After this report the sentence of the Inquisition followed in 1633 and stated: "We decree that the book, *Dialogue of Galileo Galilei,* be prohibited by public edict, and We condemn you to the prison of this Holy Office for a period determinable at Our pleasure, and by way of salutary penance. We order you during the next three years to recite, once a week, the seven penitential psalms, reserving to Ourselves the power of moderating, commuting, or taking away the whole or part of the said punishment or penance." In response to this sentence, which is still regarded as valid by the Catholic Church, Galileo was forced to abjure and declared: "I abjure, curse and detest the said errors and heresies, and generally every error and heresy contrary to the said Holy Church, and I swear that I will never more in future say, or assert anything, verbally or in writing, which may give rise to a similar suspicion of me..."

This is a manuscript copy of the sentence against Galileo and his abjuration. With the exception of the punctuation and in the spelling of a few words, this manuscript is identical to that published by Favaro in 1907.

The file also contains the following documents related to Galileo:

Giovanni Jacopo Boccardi. Five autograph letters dated 1637-1639.

Letter by Luca Valerio to Galileo dated 24.9.1610, published by Favaro in *Le Opere...* cit., vol. 10, p. 434.

Letter by Iacopo Soldani to Galileo dated 29.7.1636, published by Favaro in *Le Opere...,* cit., vol. 18, p. 435.

Autograph letter by Francesco Noailles to Galileo dated 1.1.1638, published after a copy by Favaro in *Le Opere*... cit, vol. 17, p. 246.

Two letters by Vincenzo Galileo (1606-49), one of which is addressed to Giovanni Nardi, dated 14.12.1645 and 27.2.1648.

Letter by the disciple of Galileo, Vincenzo Ranieri (?-1648) dated 25.9.1647 and another copy of the sentence against Galileo and his abjuration.

Letter by the astronomer Angelo Secchi (1818-1867) on Galileo.

Seven letters by Filippo Salviati (?-1614) written between 1565 and 1566.

Letter by the disciple of Galileo, Francesco Rinuccini, dated 2.4.1633.

Letter by Tommaso Rinuccini dated 30.4.1656.

Letter by the Cardinal Francesco dal Monte (1549-1626) to Paolo Vinta dated 19.8.1600.

Letter by Giovanni de Medici dated 1.11.1608.

Letter by Ferdinando II de Medici (1610-1670) dated 12.3.1640.

Letter by Giovanni Carlo de Medici dated 6.2.1642.

Cosimo II (de Medici) (1590-1621). Official document on vellum signed by Cosimo and dated 2.5.1609.

Letter by the disciple of Galileo, Raffaello Magiotti (1597-1658) addressed to Giambattista Doni and dated 21.9.1647.

Manuscript copy of Galileo's "Esplicazione di alcuni versetti del Salmo XVIII di David. Di Galileo-Galilei". Dimensions: 29.3 x 20.1 cms. 4 leaves. Undated. In the manuscript there is a note saying that it was published in 1777 but I have not been able to find it in *Le Opere*... cit. under the title given in this copy.

Letter by Gerolamo Galilei (1604-1670) dated 12.5.1626.

Two Letters by Francesco Galilei dated 30.8.1630 and 9.2.1635.

An autograph by Elisabetta Galilei (daughter of Galileo) dated 8.8.1639.

Two letters by Antonio and Ottavio Galilei dated 1628.

Letter by the disciple of Galileo, Carlo Dati (1619-1676), dated 1636.

Letter by the disciple of Galileo, Giovanni Battista Ciampoli (1589-1643), addressed to Mons. Caffarelli and dated 28.7.1638.

Letter by the relative of Galileo, Alessandro Bocchineri, dated 23.7.1630.
Undated letter by Cardinal Roberto Bellarmino (1542-1621).
Letter by Cesare Baronio (1538-1607) dated 5.1.1590.
Two letters by Cardinal Ottavio Bandini (1558-1629) dated 19.8.1593 and 25.4.1610.
Letter by Cardinal Francesco Barberini (1597-1673) dated 3.7.1641.
Letter by Filippo Arrighetti (1582-1662) dated 26.8.1630.
Manuscript of sixteen leaves bearing the title "Scritture sopra la Disputa avuta da N.N. con il Sig: Andrea Arrighetti".
List of specific gravities of metals and of different weights and measures. (Manuscript of two leaves written in Italian).
In addition to the letters by Torricelli and Borelli (see items No. 27 and No. 28), there is a letter dated 28.12.1654 by Vincenzo Viviani, another disciple of Galileo. The letter is kept in Waller 650 A:2 (file "Viviani").

23.

[GALILEI, GALILEO, (1564-1642)]

MANUSCRIPT	Italian. Unpublished?
TITLE	Il sistema [di] Copernico illustrato dal Galileo.Canzone.
PLACE & DATE	[Florence? Second half of the XVIIth century].
DIMENSIONS	21.5 x 14.9 cms.
FOLIATION	4 leaves.
SHELFMARK	Waller 650 A:2.
CONTENTS & TRANSCRIPTION	

Poem written after the death of Galileo containing an allegory of Truth dictating to Galileo the system of the universe and encouraging him to fight the obscurity of ignorance. In it the Copernican system of the universe is described and the movement of the earth asserted. Truth also announces to Galileo that his work would meet the opposition of ignorance but no mention of the Catholic Church is made. Because of damage to the manuscript some words are not legible and have been indicated by the sign [...], whereas those added by the editor have been enclosed in brackets. The whole text is:

Il Sistema [di] Copernico illustrato dal Galileo.

Canzone.

Se mai dell'Arno alla feconda riva
Oggi volgesse il piede
Quel sì famoso veglio di Stagira
Oh come acceso d'ira
Mirerebbe raminga e fuggitiva
(Per tanti lustri già Donna e Signora)
La sua filosofia!
Dell'Ignoranza il vel, che fino ad ora
Baldanzosa regnò, rotto vedria:
Poiche l'Errore invan [tenta], e presume
Danno recar di Veritade al Lume.

Già spiegava nel ciel l'umide [om]brose
Piume la notte oscura,
E fugava dal sen le pene, e i mali
De' miseri mortali
E posto il fren delle create cose
Parea dormir Natura;
Pure il gran Galileo forse presago
Non chiude al sonno i lumi:
In un manto gli appar semplice e vago
Una che al certo vien dà sommi Numi,
Che in esso lui tenendo gli occhi intenti
Facegli udire il suon di questi accenti:

Io figlia son di facitor superno,
Che a lui mi siedo a lato
Sopra il lucente ed immortal suo trono
La veritade io sono;
A te, scesa nel sen del Nume Eterno,
Ne vengo [o] figlio amato
Da me prescelto a gloriose imprese.
O figlio è tempo omai
Ch'io vendichi per te le proprie offese:
Abbastanza finor soffersi, assai
Dal velo dell'Errore [orrido], e nero
Io fui celata all'Universo intero.

E fino a quanto in vergognoso oblio
Sconosciuta a' mortali
Dovrò giacerne ascosa? e fino a quanto

L'ingenito mio vanto
Ignorarsi dovrà? Svolger degg'io
Ciò che gli eterni annali
Non palesano ancora; dai sommi Dei
Del sistema del mondo
Le leggi a palesar prescelto sei
Finor sepolto in un oblio profondo:
Vanne dunque, e lo svela all'uomo cieco,
E non temer, la veritade è teco.

Narra al [missal], che la grand'ignea mole
Anzi ocean di fuoco,
Per la cui [...] e sempiterna luce
[Ivi] splende, e riluce,
E' nel centro del mondo, io dico il sole,
Che immobil nel suo loco
Sta bene ognora; oh meraviglie rare!
Vi sono altre opre belle,
In cui la mano dell'Architetto appare,
Dal moto lor chiamate erranti stelle;
Che con costante, e fissa legge intorno
Compion lor giri al portator del giorno.

Vicino al sole in quelli [spazi] immensi
Del cielo il nunzio alato
Compie il suo giro, Dio facondo, e saggio,
Cui l'Apollineo raggio
Lo fa di rado comparire ai sensi
Dell'[uman] occhio armato.
Quella che segue luminosa stella
Più lungi dal sol gira;
E' la madre di Amor Venere [...]
Che grazie, e Amori [...];
E quando sorge, o quando [va] in mare
Niuna più bella, o risplendente appare.

Quindi ne segue la Terrena mole,
Che sotto al piè ti miri,
Che in monti, in colli, in valli, in pian distinta
Frena il mare ond'è cinta;
E intorno a se medesima, e intorno al sole,
Compie i certi suoi giri.
Colei che rende il ciel di note adorno
Del sole alma sorella,

Per far suo giro al vostro globo intorno,
Della terra satellite si appella,
Che vi mostra (perciò detta riforma)
L'alta bellezza sua sotto più forme.

Del ciel le strade a contemplar tornando
In più remota parte
Con quella scabra ruvida faccia
A sua sfera s'affaccia
Quel Dio che cinge e la lorica, e il brando,
Il sanguinario Marte:
Vedessi pure rimirando altrove,
Che il sol suo lume dona
A un immenso pianeta, e questi è Giove
Intorno a cui formando una corona
Girano quattro lune argentee e belle
Le quali appellerai Medicee stelle.

Si scorge al fin la più discosta stanza
Di Saturno soggiorno
E in queste parti poichè stanchi omai
Giungon del sole i rai,
D'un vasto cerchio fiammeggiante è adorno
E in diversa distanza
Cinque il circondan sfolgoranti faci.
Ciò palesar conviene
Il mondo, e dileguar l'idee fallaci:
Molte tu soffrirai fatiche, e pene,
Ma non temer sii nell'oprar costante,
Quindi si tacque, e mi spario d'avante.

Già la [grand']opra cominciata avea
L'Immortal Galileo,
Quando dal seno del profondo oblio
Ignoranza sortio,
Chè ovunque folte tenebre spargea;
Ma alfin più non potea
Di veritade opporsi al vivo sole
Che co' suoi rai lucenti
Tutta illustrava la terraquea mole;
Quindi fremendo a passi tardi e lenti
Si nascose nel [Infero] profondo
Più non osando comparire al mondo.

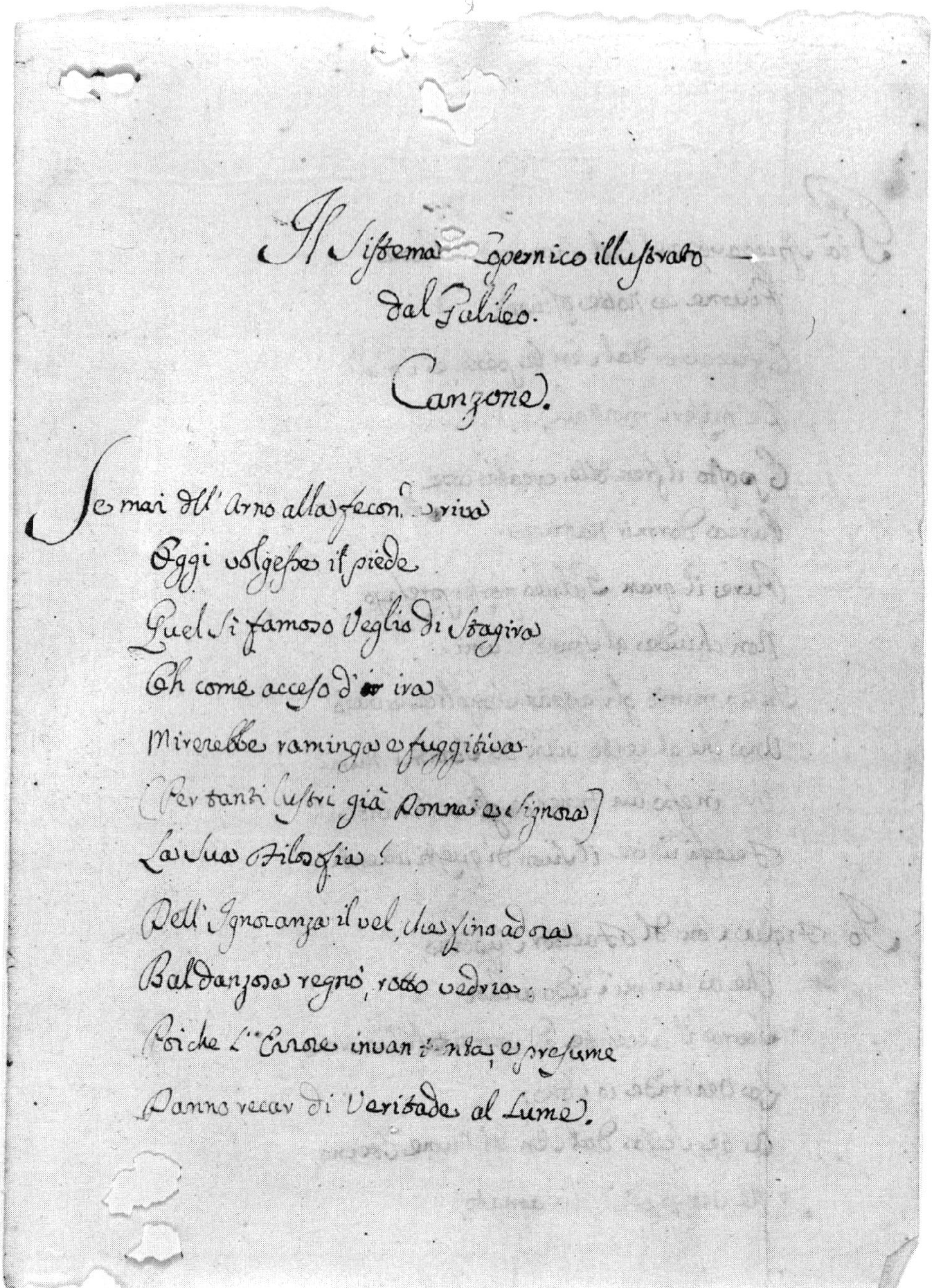
Il Sistema Copernico illustrato
dal Galileo.
Canzone.
Se mai dell'Arno alla fecon[illegible] riva
Oggi volgesse il piede
Quel sì famoso Veglio di Stagira
Oh come acceso d'ira
Mirerebbe raminga e fuggitiva
(Per tanti lustri già Donna e Signora)
La sua Filosofia [illegible]
Dell'Ignoranza il vel, che sino ad ora
Baldanzosa regnò, rotto vedria.
Poiche l'Errore invan [illegible]nta, e presume
Danno recar di Veritade al Lume.

Fig. 23.

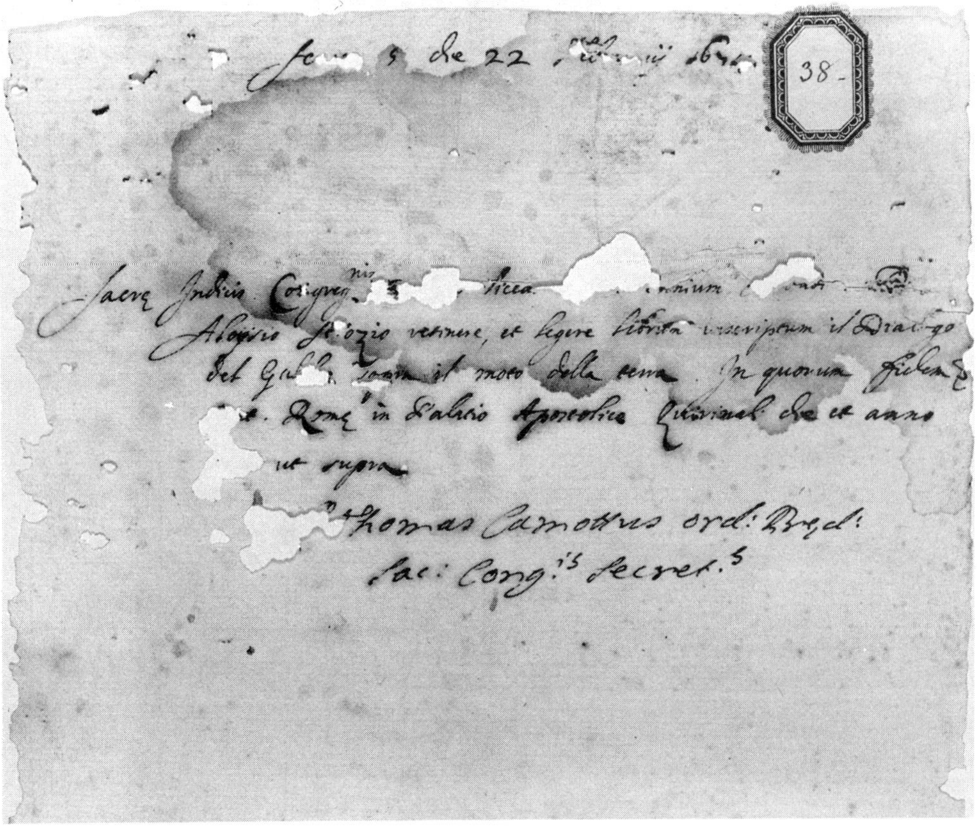
[illegible] die 22 [illegible] 16[illegible]

38.

Sacrae Indicis Congreg.nis [illegible] licentia [illegible] Aloysio Strozio retinere, et legere librum inscriptum il Dialogo del Galil. [illegible] il moto della terra. In quorum fidem [illegible]. Romae in Palatio Apostolico Quirinali die et anno ut supra.

Fr. Thomas Camottus ord. Praed. Sac. Cong.is Secret.s

24.

CAMOTTUS, THOMAS

LICENCE	Licence to read Galileo's *Dialogo sopra i due massimi sistemi del mondo* to Luigi Strozzi.
PLACE & DATE	[Florence?], 22.2.1674.
DIMENSIONS	27 x 19.1 cms.
FOLIATION	1 leaf.
SHELFMARK	Waller 650 A:2.
NOTE	This valuable document provides evidence that as late as 1674 the reading of Galileo's main work was still strictly controlled by the Church. However, this licence given to Strozzi shows that the censorship of the Church allowed of some exceptions.

25.

[RÜMMELIN, JOHANNES (1583-1632)]

MANUSCRIPT Latin. Published.

TITLE Catoptrum microcosmicum suis aere incisis visionibus splendens cum historia, et Pinace, de novo Prodit...

PLACE & DATE Added in pencil, Ulm 1639.

DIMENSIONS 45.6 x 33.1 cms.

FOLIATION 13 leaves.

ILLUSTRATED

SHELFMARK Waller 653 F:2.

HISTORICAL NOTE

This seems to be the preparatory version for the 1639 edition of Rümmelin's famous anatomical work. The anatomical illustrations with movable flaps (which were introduced to illustrate treatises of anatomy at the beginning of the XVIth century), ensured Rümmelin's work a resounding success. The three anatomical plates were composed of more than one hundred moveable flaps representing the internal constitution of the human body. According to O. Hagelin (*Rare and Important Medical Books in the Library of the Swedish Society of Medicine,* Stockholm, 1989, p. 59), the plates and the manuscript were published by Rümmelin's friends without his knowledge. If this was the case it seems more than likely that this was the source manuscript of the edition of 1639. The frontispiece, the text, the foliation, and the disposition of the plates in fact correspond to the edition of 1639.

The illustrations adorning Rümmelin's work were a mixture of allegorical and descriptive representations of the human anatomy. In comparison with the iconographical apparatus of the anatomical treatises published in the previous century by Vesalius and Estienne, one cannot fail to notice that the scientific accuracy of their observations has been replaced by Rümmelin with a symbolical representation of the human body and that accuracy of anatomical detail is no longer as important as it was for the anatomists of the Renaissance. In Rümmelin's work in fact metaphysical and alchemical beliefs played an important role and the notion of experience was mediated by speculative beliefs. In addition to this manuscript version, Waller collected ten different editions of Rümmelin's work (see *Bibliotheca Walleriana,* nos. 7880-89).

26.

VOSSIUS, ISAAC (1618-1689)

LETTER — Latin. Unpublished. The file also contains one more letter dated October 1668 in Dutch from Hamburg and a receipt dated Uppsala 1.2.1654, the year in which Vossius was in Uppsala with the task of enlarging the library of Queen Christina of Sweden.

PLACE & DATE — [Leiden], 16.8.1659.

ADDRESSEE — D. Huet.

DIMENSIONS — 20.2 x 16 cms.

FOLIATION — 4 leaves.

SHELFMARK — Waller 649 F:2.

CONTENTS

Scientific letter in which Vossius discusses Kepler's theory of lunar refraction and, more generally, the theory of refraction of bodies. In the letter Vossius also criticized the philosophy of Descartes and adds: "Ingenium vero laudare non possum, si huius magis quam veritatis habeatur ratio". Vossius, who met Descartes in 1649 at the court of Queen Christina in Stockholm, never forgave the French philosopher for his criticism of his erudition and his belief in the superiority of philosophy over philology. On this episode, see C. Adam, *Vie & Oeuvres de Descartes*, (Paris, 1910), p. 542.

27.

TORRICELLI, EVANGELISTA (1608-1647)

LETTER — Italian and Latin. Unpublished. The same file also contains a non-autograph copy of a letter by Torricelli to Raffaello Magiotti dated 4.12.1643 which was published in E. Torricelli, *Opere*, vol. 3, (Faenza, 1919), pp. 150-154.

PLACE & DATE — Florence, 17.1.1645.

ADDRESSEE — Michelangelo Ricci.

DIMENSIONS — 27.5 x 20.3 cms.

FOLIATION — 2 leaves.

SHELFMARK — Waller Delikatesskaps 1.

CONTENTS

Letter in which Torricelli claims priority in demonstrating that the line of a parabola was equal to the spiral of Archimedes and communicates the demonstration to Ricci. In this connection

Torricelli writes: "avrei caro sapere dal P. Mersenne chi sia stato l'autor in Francia che ha detto di mostrar la linea curva parabolica eguale ad una spirale Archimedea; io ho conseguito ancor'io tal dimostrazione".

HISTORICAL NOTE

While writing this letter Torricelli, probably one of the most prominent disciples of Galileo, did not know that the French mathematician who performed the demonstration was Fermat. It is significant and typical of the period that in order to claim priority for his scientific discovery, Torricelli asked Ricci to show the letter or a copy of it to Mersenne, Magiotti and Nardi. To vindicate priority in the case of a scientific discovery it was common to make several copies of a letter and to put them into circulation. However, in response to Torricelli's claim, Michelangelo Ricci told him that Fermat's demonstration differed substantially from his own, for which reason he had nothing to fear (see E. Torricelli, *Opere*, vol. 3, Faenza, 1919, p. 256 and ff.)

28.

BORELLI, GIOVANNI ALFONSO (1608-1679)

LETTER Latin. Unpublished.
PLACE & DATE Messina, 6.12.1670.
ADDRESSEE John Wallis in Oxford.
DIMENSIONS 29.2 x 20.1 cms.
FOLIATION 1 leaf.
SHELFMARK Waller 649 E:1.
TRANSCRIPTION (Borelli's e̱ caudata has been transcribed with a̲e̲)

Clariss.° Doctissimoq. viro D. Johanni Wallisio in Academia Oxoniensi matheseos professori Saviliano.

Jo. Alphons. Borellus. S

Accepi tandem (vir clarissime) diu desideratam tuam epistolam quae diuturnam moram longe maioribus beneficiis compensavit; quippequae dona amplissima et amicitiam pretiosam clarissimorum virorum D. Boylei, et D. Collinsii una cum tuis libris attulit, pro qua beneficentia summopere me tibi obstrictum profiteor. Doleo tamen libris Clar. Boyle frui non posse cum Anglici idiomatis prorsus ignarus sim; verum non despero auxilio anglorum hic degentium posse licet tardius eorum aliquam interpretationem consequi: interim ex titulis, ex figuris et ex

genio authoris ex operibus eius latinis praecognito coniecisse me puto quid in hisce contineatur. quae non videntur recedere ab ea philosophia, quae valde mihi arridet. Tua vero opera exprimere non possum quantopere me delectarunt, ob ingenii acumen et perspicaciam, inventionum copiam et iudicium adaequatum quo veriora a dubiis seligere conaris. dolet tamen q[u]od etiam apud vos non desint scioli qui livore, aut ignorantia veritate[m in]sectantes doctos distrahant, et in apologiis conficiendis bonas horas terere inutiliter cogant: hoc malum apud nos minus grave videtur, quia assiduè nos vexat. prodiit nuper ex Gallia dialogus religiosi cuiusdam e societate Iesu Honorati Fabri qui integrum meum opus de vi percussionis superbissimo contemp[t]u reiicit. huic profecto nil respondissem, nisi forte turpissima nota me affecisset. ergo coactus brevissime ei respondeo in prohemio opusculi de nupero AEtnae incendio quod sub prelo sudat, et quam primum ad vos mittam. Circa libros quos D. Boyle a me petit, video vos a titulis moveri dissitorum scriptorum, sicuti nos sepe decipimur. Huius generis sunt libri in scheda vestra adnotati, aliqui eorum sunt adeo pueriles ut contemptus ipsae [sic] raritatem eorum effecerit. Exercitationes mechanicae Marchetti non respondebunt tuae expectationi. tractatus Rinaldini de curvis mediceis non est editus, sicut nec tractatus eiusdem argumenti Doctiss. Riccii: qui negociis ecclesiasticis Romae implicitus mathematicos otiari sinit. Grimbergeri de luce et refractionibus opus postumum certe non extat. sed eius vice prodiit Grimaldi opus eodem titulo. suspicor a nominis similitudine vos decipi. Hodiernae [sic] opera non sunt tanti facienda, is fuit parum matheseos peritus, edidit aliqua astronomica. Amicis tamen Romae, Pisis, Florentiae et Bononiae scripsi ut predictos libros undique perquirerent. Dumque eos expecto se obtulit repentina huius navigii occasio. non eam censui omittendam, mitto hinc qui apud me extant libros, reliquos liburni forsam transmittent amici. interea, vir clariss.[e] me amare perge et clariss. D. Boyleum et D. Collinsium salutes rogo. Vale.
Messane 6. Decembris 1670.

HISTORICAL NOTE

Interesting letter by the disciple of Galileo, Borelli, to the English mathematician Wallis, in which Borelli informs him of the latest publications in physics and mathematics printed in Italy and gives his opinion (not quite favorable) on some of them. Borelli also discusses his controversy with the Jesuit Fabri.

Fig. 25 a: *Frontispiece.*

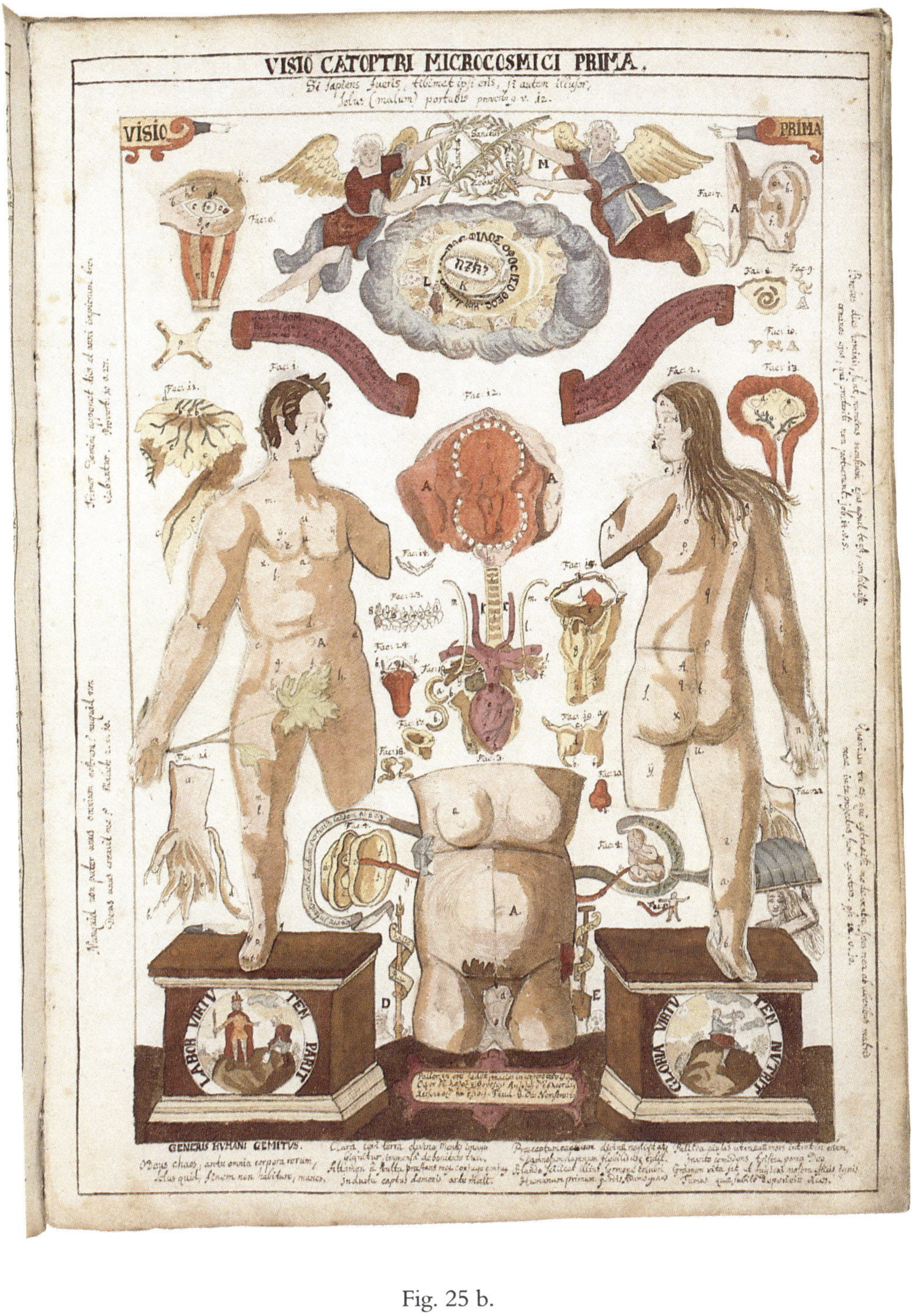
VISIO CATOPTRI MICROCOSMICI PRIMA.
VISIO
PRIMA
LABOR VIRTU TEM PARIT
GLORIA VIRTU TEM NUTRIT
GENERIS HUMANI GEMITUS.

Fig. 25 b.

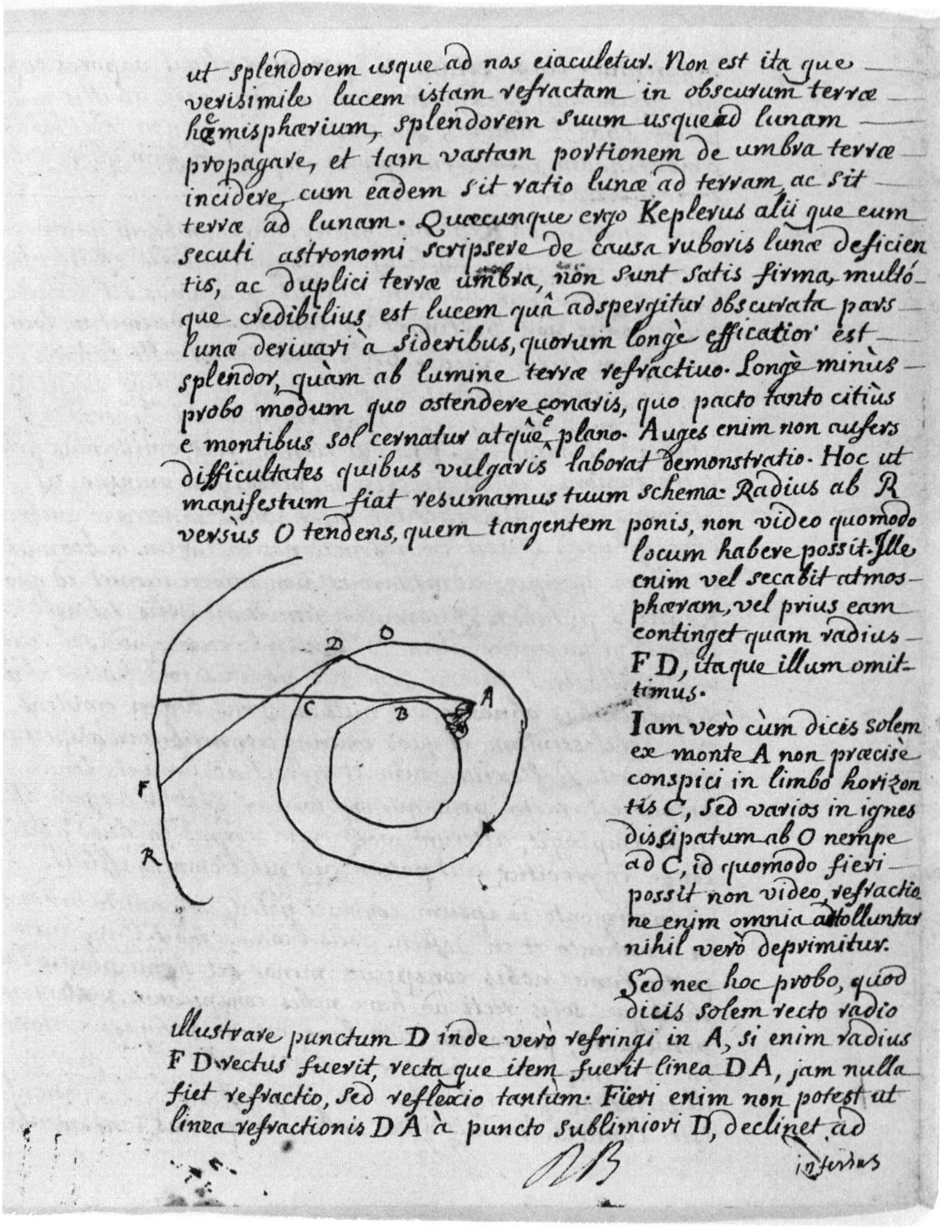

ut splendorem usque ad nos eiaculetur. Non est itaque verisimile lucem istam refractam in obscurum terrae hemisphaerium, splendorem suum usque ad lunam propagare, et tam vastam portionem de umbra terrae incidere, cum eadem sit ratio lunae ad terram ac sit terrae ad lunam. Quaecunque ergo Keplerus aliique eum secuti astronomi scripsere de causa ruboris lunae deficientis, ac duplici terrae umbra, non sunt satis firma, multóque credibilius est lucem quâ adspergitur obscurata pars lunae derivari à sideribus, quorum longè efficatior est splendor, quàm ab lumine terrae refractivo. Longè minùs probo modum quo ostendere conaris, quo pacto tanto citiùs e montibus sol cernatur atque e plano. Auges enim non aufers difficultates quibus vulgaris laborat demonstratio. Hoc ut manifestum fiat resumamus tuum schema: Radius ab R versus O tendens, quem tangentem ponis, non video quomodo locum habere possit. Ille enim vel secabit atmosphaeram, vel prius eam continget quam radius F D, itaque illum omittimus.

Iam verò cùm dicis solem ex monte A non praecisè conspici in limbo horizontis C, sed varios in ignes dissipatum ab O nempe ad C, id quomodo fieri possit non video, refractione enim omnia attolluntur nihil verò deprimitur.

Sed nec hoc probo, quod dicis solem recto radio illustrare punctum D inde verò refringi in A, si enim radius F D rectus fuerit, rectaque item fuerit linea D A, jam nulla fiet refractio, sed reflexio tantum. Fieri enim non potest ut linea refractionis D A à puncto subliniori D, declinet ad

inferius

Fig. 26.

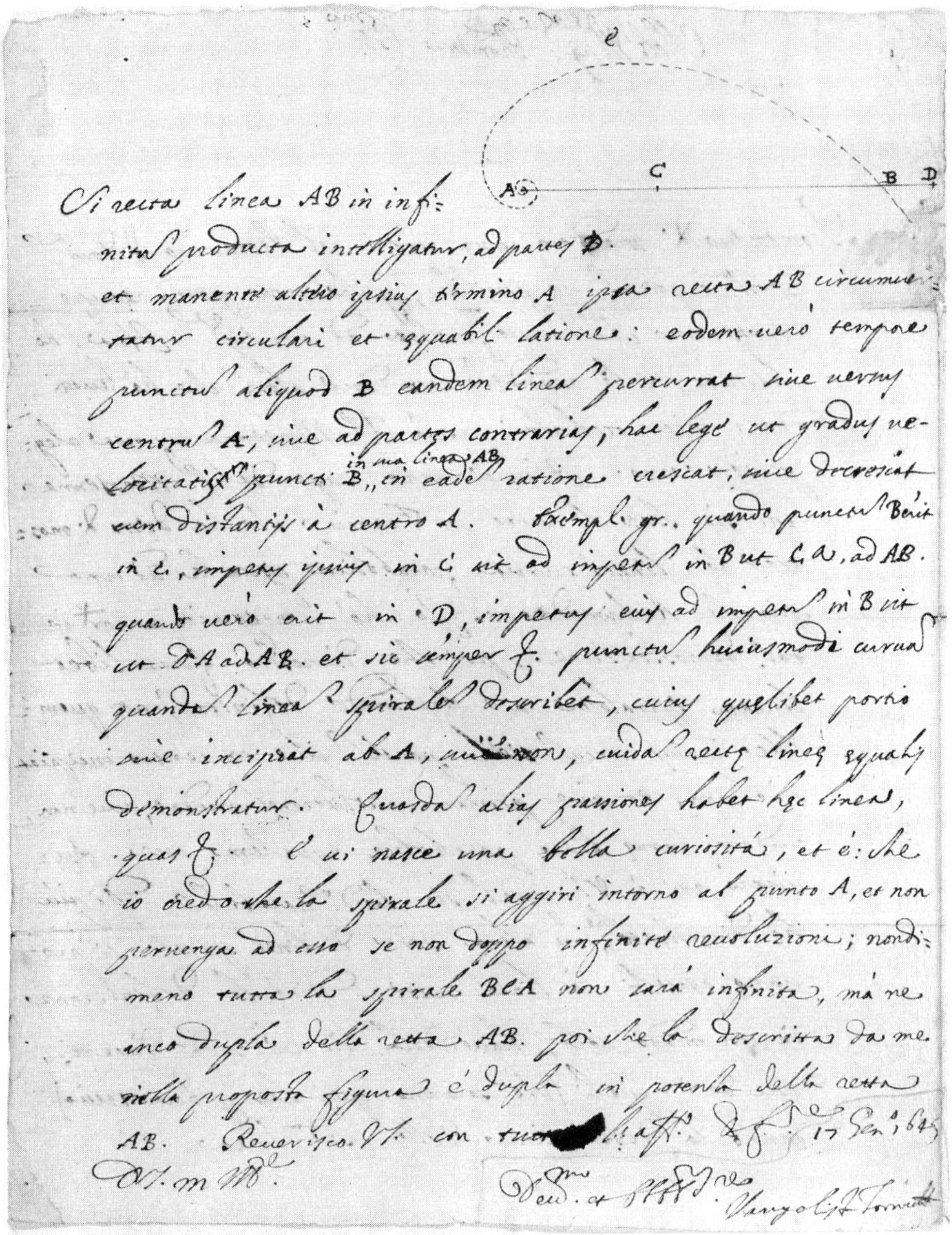

Si recta linea AB in infi-
nitum producta intelligatur, ad partes D.
et manente altero ipsius termino A ipsa recta AB circumve-
tatur circulari et aequabili latione: eodem vero tempore
punctum aliquod B eandem lineam percurrat sive versus
centrum A, sive ad partes contrarias, hac lege ut gradus ve-
locitatis puncti B, in una linea AB, in eadem ratione crescat, sive decrescat
cum distantiis à centro A. Exempl. gr. quando punctum B erit
in C, impetus ipsius in C sit ad impetum in B ut CA, ad AB.
quando vero erit in D, impetus eius ad impetum in B sit
ut DA ad AB. et sic semper &c. punctum huiusmodi curvam
quandam lineam spiralem describet, cuius quaelibet portio
sive incipiat ab A, sive non, cuidam rectae lineae aequalis
demonstratur. Quasdam alias passiones habet haec linea,
quas &c. E vi nasce una bella curiosità, et è: che
io credo che la spirale si aggiri intorno al punto A, et non
pervenga ad esso se non doppo infinite revoluzioni; nondi-
meno tutta la spirale BCA non sarà infinita, ma ne
anco dupla della retta AB. poi che la descritta da me
nella proposta figura è dupla in potenza della retta
AB. Reverisco V.S. con tutto [illegible] l'affetto. Di F.ze 17 Gen.o 1645
V.S. m. Ill.re

Dev.mo et Oblig.mo Vangelista Torricelli

Fig. 27.

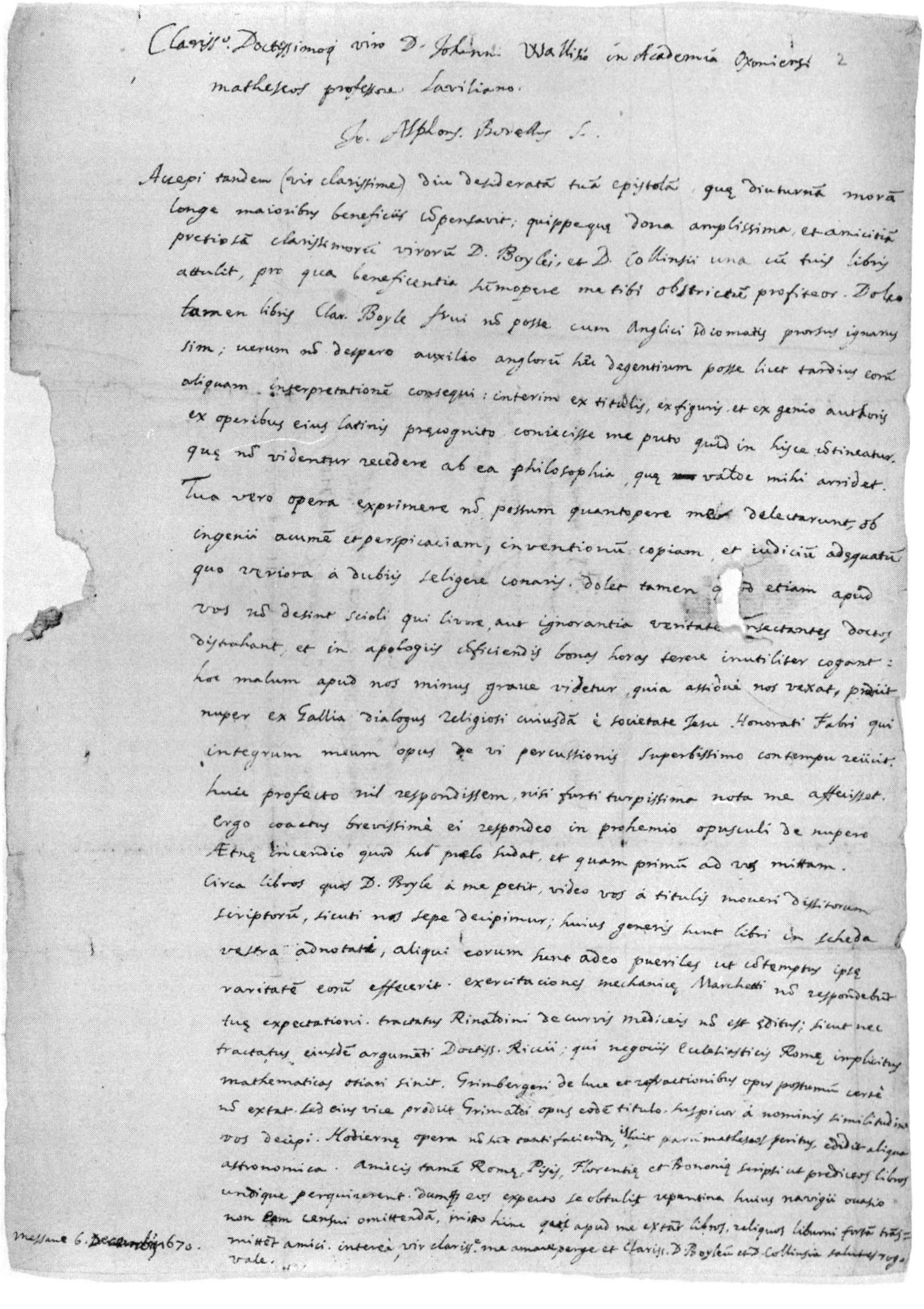

Clariss.o Doctissimoq viro D. Johanni Wallisio in Academia Oxoniensi 2
mathiseos professore Saviliano.

Jo. Alphons. Borelius S.

Accepi tandem (vir clarissime) diu desideratã tuã epistolã, quę diuturnã morã longe maioribus beneficiis compensavit; quippe quę dona amplissima, et amicitiã pretiosã clarissimorũ virorũ D. Boylei, et D. Collinsii una cũ tuis libris attulit, pro qua beneficentia summopere me tibi obstrictũ profiteor. Doleo tamen libris Clar. Boyle frui nõ posse cum Anglici idiomatis prorsus ignarus sim; verum nõ despero auxilio anglorũ hic degentium posse licet tardius eorũ aliquam interpretationẽ consequi: interim ex titulis, ex figuris et ex genio authoris ex operibus eius latinis praecognito coniecisse me puto quod in hisce continentur quę nõ videntur recedere ab ea philosophia, quę valde mihi arridet. Tua vero opera exprimere nõ possum quantopere mihi delectarunt, ob ingenii acumẽ et perspicaciam, inventionũ copiam, et iudiciũ adaequatũ quo veriora à dubiis seligere conaris. Doleo tamen quod etiam apud vos nõ desint scioli qui livore, aut ignorantia veritatem insectantes doctos distrahant, et in apologiis conficiendis bonas horas terere inutiliter cogant: hoc malum apud nos minus grave videtur, quia assiduè nos vexat, prodiit nuper ex Gallia dialogus religiosi cuiusdã è societate Jesu Honorati Fabri qui integrum meum opus de vi percussionis superbissimo contemptu reiicit. huic profecto nil respondissem, nisi furti turpissima nota me affecisset. Ergo coactus brevissimè ei respondeo in prohemio opusculi de nupero Aetnę incendio quod sub praelo sudat, et quam primũ ad vos mittam. Circa libros quos D. Boyle à me petit, video vos à titulis moveri dissitorum scriptorũ, sicuti nos saepe decipimur; huius generis sunt libri in scheda vestra adnotati, aliqui eorum sunt adeo pueriles ut contemptus ipse raritatẽ eorũ effecerit. exercitationes mechanicę Marchetti nõ respondebit tuę expectationi. tractatus Rinaldini de curvis medicęis nõ est editus; sicut nec tractatus eiusdẽ argumẽti Doctiss. Ricii; qui negotiis ecclesiasticis Romę implicitus mathematicas otiari sinit. Grimbergeri de luce et refractionibus opus posthumũ certè nõ extat. sed eius vice prodiit Grimaldi opus eodẽ titulo. suspicor à nominis similitudine vos decipi. Hodiernę opera nõ sunt satisfacienda, [illegible] parum mathiseos peritus, edidit aliqua astronomica. Amicis tamẽ Romę, Pisis, Florentię et Bononię scripsi ut praedictos libros undique perquirerent. dumque eos expecto se obtulit repentina huius navigii occasio non eam censui omittendã, mitto hinc quos apud me extant libros, reliquos liburni forsan transmittet amici. interea vir clariss.e me amare perge et Clariss. D. Boyleũ et D. Collinsiũ salutes rogo. vale.

Messanę 6. Decembris 1670.

Fig. 28.

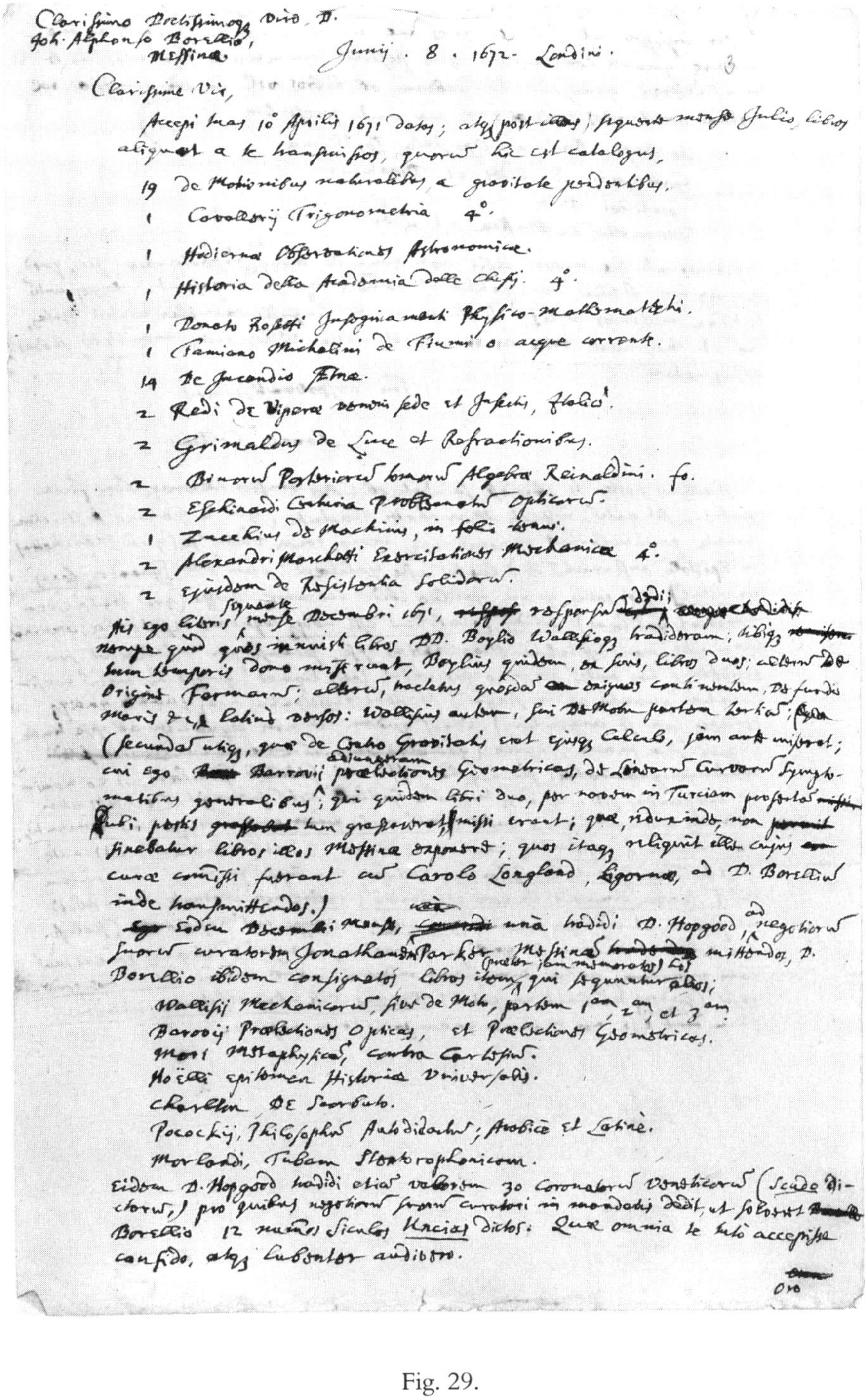

Clarissimo Doctissimoque Viro D.
Joh. Alphonso Borellio,
Messinae.

Junij. 8. 1672. Londini.

Clarissime Vir,

Accepi tuas 10° Aprilis 1671 datas; atque post illas, sequente mense Julio, libros aliquot a te transmissos, quorum hic est catalogus,

19 De Motionibus naturalibus, a gravitate pendentibus.
1 Cavallerij Trigonometria 4°.
1 Hodiernae Observationes Astronomicae.
1 Historia della Academia delle Chiesi. 4°.
1 Donato Rossetti Insegnamenti Physico-mathematichi.
1 Famiano Michelini de Fiumi o acque correnti.
14 De Incendio Aetnae.
2 Redi de Viperae veneni sede et Insectis, Italicè.
2 Grimaldus de Luce et Refractionibus.
2 Binorum Posteriorum tomorum Algebrae Reinaldini. fo.
2 Eschinardi Centuria Problematum Opticorum.
1 Zucchius de Machinis, in folio [illegible].
2 Alexandri Marchetti Exercitationes Mechanicae 4°.
2 Ejusdem de Resistentia Solidorum.

His ego libris sequente mense Decembri 1671, responsum dedi; quos misisti libros DD. Boylio Wallisioque tradideram; tibique [illegible] temporis dono miserunt, Boylius quidem, ex suis, libros duos; alterum De Origine Formarum; alterum, tractatus quosdam origines continentem, De fundis maris etc. Latinè versos: Wallisius autem, sui De Motu partem tertiam; (secundam utique, quae de Centro Gravitatis erat ejusque Calculo, jam ante miserat; cui ego Barrovii praelectiones Geometricas, de Linearum Curvarum Symptomatibus generalibus adjunxeram; qui quidem libri duo, per nostrum in Turciam profectum [illegible] tibi, postis tum grassantis missi erant; quae, ridere inde, non sinebatur libros illos Messinae exponere; quos itaque reliquit ille, cujus curae commissi fuerant cum Carolo Longland, Ligornae, ad D. Borellium inde transmittendos.)

Eodem Decembri mense, una tradidi D. Hopgood, ad negotiorum suorum curatorem Jonathanem Parker Messinae mittendos, D. Borellio ibidem consignatos libros itidem (praeter jam memoratos) hos qui sequuntur alios;

Wallisij Mechanicorum, sive de Motu, partem 1am, 2am, et 3am.
Barovij Praelectiones Opticas, et Praelectiones Geometricas.
Mori Metaphysicam, contra Cartesium.
Hoëlli epitomen Historiae Universalis.
Charelton De Scorbuto.
Pocockij, Philosophum Autodidactum; Arabicè et Latinè.
Morlandi, Tubam Stentorophonicam.

Eidem D. Hopgood tradidi etiam valorem 30 Coronatorum Venetiorum (Scudo dictorum,) pro quibus negotiorum suorum curatori in mandatis dedit, ut solveret Borellio 12 nummos Siculos Uncias dictos: Quae omnia te tuto accepisse confido, atque lubenter audivero.

Oro

Fig. 29.

29.

COLLINS, JOHN (1625-1683)

LETTER Latin. Unpublished.
PLACE & DATE London, 8.6.1672.
ADDRESSEE G.A.Borelli.
DIMENSIONS 29 x 19 cms.
FOLIATION 1 leaf.
SHELFMARK Waller Delikatesskaps 1.
TRANSCRIPTION

Clarissimo Doctissimoque viro; D.
Joh. Alfonso Borellio
Messinae Junij. 8. 1672 Londinii
Clarissime vir,

Accepi tuas 10° Aprilis 1671 datas; atque (post illas) sequente mense Julio, libros aliquot a te transmissos quorum hic est catalogus,

19 De motionibus naturalibus a gravitate pendentibus.
1 Cavallerij Trigonometria 4°.
1 Hodiernae Observationes Astronomicae.
1 Historia della Academia dalle [Classij]. 4°.
1 Donato Rosetti Insegniamenti Physico-Mathematichi.
1 Famiano Michelini de Fiumi e acque corrente.
14 De Incendio AEtnae.
2 Redi de Vipera veneni sede et Insectis, Italico.
2 Grimaldus de Luce et Refractionibus.
2 Binorum-Posteriorum tomorum Algebrae Reinaldini. fo.
2 Eschinardi Centuria Problematum Opticorum.
1 Zucchius de Machinis, in folio tenui.
2 Alexandri Marchetti Exercitationes Mechanicae 4°.
2 Ejusdem de Resistentia Solidorum.

His ego libris sequente mense Decembri 1671, responsum dedi, nempe quod quos innuisti libros D.D. Boylio Wallisioque tradideram, tibique tum temporis dono miserunt Boylius quidem, ex suis, libros duos; alterum De Origine Formarum; alterum tractatus quosdam exiguos continentem, De fundis maris etc latine versos: Wallisius autem, sui De Motu partem tertiam (secundam utique, quae de Centro Gravitatis erat ejusque Calculo, jam ante miserat; cui ego, Barrovii adiunxeram praelectiones Geometricas, de Linearum Curvarum Symtomatibus generalibus, qui quidem libri duo, per navem in Turciam profectam ubi pestis tum grassaverat [sic] missi erant; quae, redux inde, non sinebatur libros illos Messinae exponere, quos itaque reliquit ille cujus curae commissi fuerant cum Carolo Longland, Ligornae, ad D. Borellium inde transmittendos).

Eodem Decembri mense una tradidi D. Hopgood, ad negotiorum suorum curatorem Jonathanem Parker Messinam mittendos, D. Borellio ibidem consignatos libros item (praeter iam memoratos) qui sequuntur alios;

Wallisii Mechanicorum, sive De Motu, partem 1^{am}, 2^{am}, et 3^{am}
Barovii Praelectiones Opticas, et Praelectiones Geometricas.
Mori Metaphysicam, contra Cartesium.
Hoëlli Epitomen Historiae Universalis.
Charlton, De Scorbuto.
Pocockii, Philosophum Autodidactum; Arabice et Latine.
Morlandi, Tubam Stentorophonicam.
Eidem D. Hopgood tradidi etiam valorem 30 coronarum veneticarum (Scudo dictarum) pro quibus negotiorum suorum curatori in mandatis dedit, ut solveret Borellio 12 nummos Siculos Uncias dictos: Quae omnia te tuto accepisse confido, atque lubenter audivero.

Oro insuper ut Librorum quos huc misisti computum digneris ad me mittere, quanti constant, quo possim numeratis pecunijs ad te transmittere, quicquid tibi debeam ob libros vel hactenus missos, vel posthac mittendos; quos iam peto sunt praesertim hi;
Jo. Bapt. Baliani Opuscula Posthuma.
Gottignij Euclides, et Dioptrica.
Galilaei Vita.
Mengolus, De Musica et Quadratura Circuli.

Prodierunt hic nuper libri novi nonnulli, aliique etiamnum expectantur, sed, quod iam agitur Bellum, haud patitur ut commode transmitti possint. Siquos autem posthac mittendos velis, si plusquam ex singulis exemplar unicum petas, oro ut insinuare digneris quot missa velis; tuisque mandatis lubens obsequetur

Tui observantissimus
Johannes Collins

Wallisius noster te officiose salutat, ob libros missos caeteraqua beneficia gratus. Ait autem; inspecto Marchetti Tractatu, (qui a

Galilaeo strictim tradita particularius exponit, neque male,) non videre se [quod Marchettus in Epistola insinuatum it] lapsum esse Galilaeum: cum illud (quandoque bonus dormitat) sibi potius quam Galilaeo ibidem debeatur. Quippe (quod legitime demonstrat Galilaeus) solidum parabolicum esse ubique aequalis resistentiae; omnino verum esse, non diffitetur Marchettus ipse, secluso per intellectum suo pondere, (hoc autem omnino faciendum, satis liquet; quippe hic tantum consideratur, quanta sit in singulis partibus resistentia; non, quanto quaeque pondere per se urgeatur.) verum quidem est, non aequaliter ab ipsa trabe urgeri sua puncta singula sed neque hoc affirmat Galilaeus, ut eo nomine culpandus sit. Quodque tanquam a Galilaeo dictum subiungit, Fieri posse ut in posterum utamur ligneis trabibus, tertia molis parte adeoque ponderis imminutis, nihil imminuta resistentia; etiam vere dicitur, sed et (Gailaeum decet) caute. Si dixisset, nihil mutata resistentia, esset forte cur lapsum putemus: cum autem ait [senza diminuir la loro gagliardia] robore non diminuto, tantum abest ut peccet, ut addidisse potuerit, imo aucto. Quippe cum resistentia (secluso suo pondere) eadem sit quae prius, robori suo minus detrahet pondus suum, iam minutum, quam dum fuerat integrum. Quod autem solidum hoc ponderosum (pondere suo simul computato) aequalis esse ubivis resistentiae iactitaverit, (dum dixerit esse, non minoris quam prius), non reperio.

HISTORICAL NOTE

Letter in which the English mathematician John Collins acknowledges the receipt of a large number of Italian scientific books (some of which are in several copies) and comments on some of them. This letter testifies to the cultural importance of the lively exchange of books during the XVIIth century. Collins also discusses in detail Marchetti's work on the resistance of bodies.

Fig. 30.

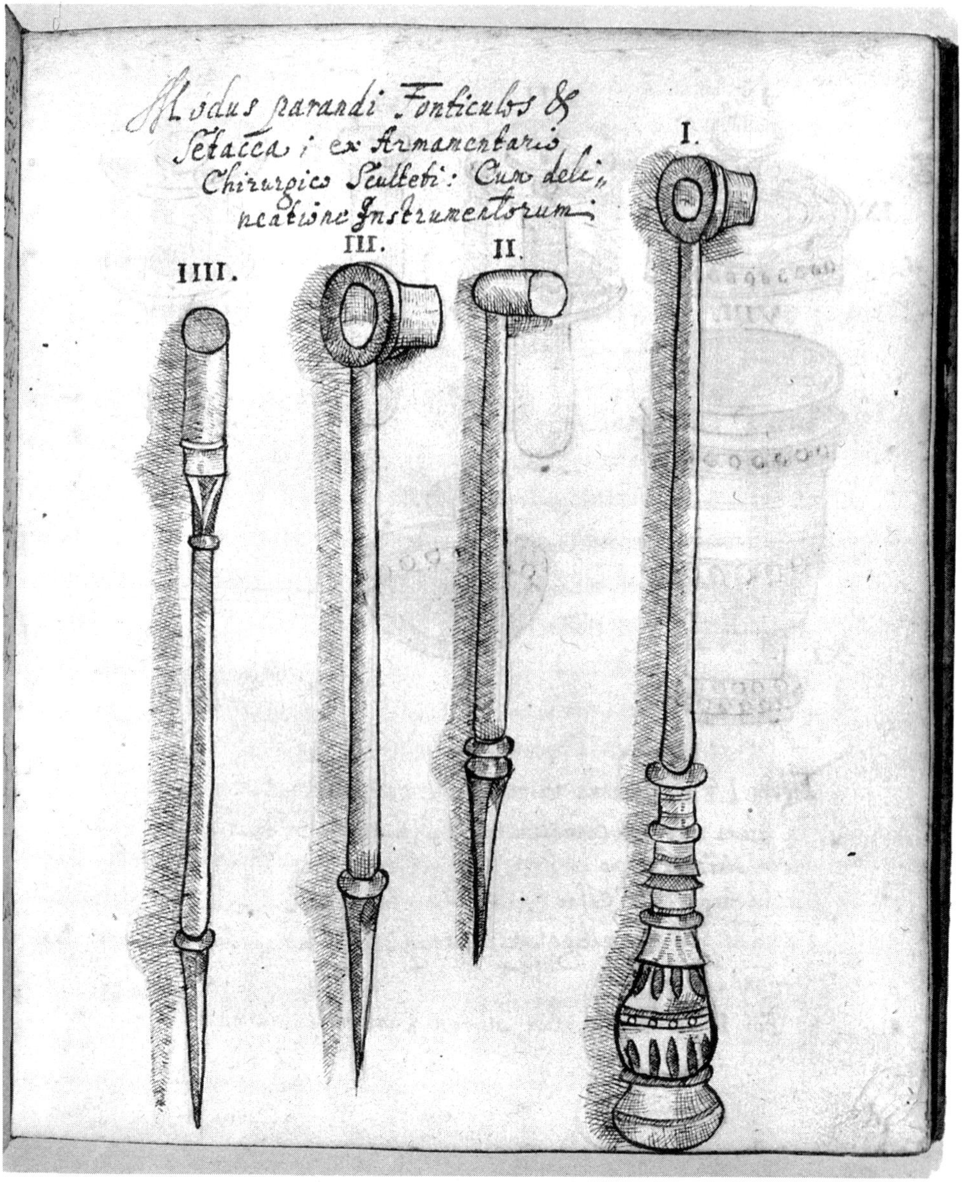

Fig. 31: *Surgical instruments*.

30.

[REDI, FRANCESCO (1626-1698)]

MEDAL — Described by Storer (see item No. 9) no 2975 as follows:

"Ob. FRANCISCVS. REDI. PATRITIVS. ARETINVS. Bust to right. Below M. SOLD[anus]. 1684.

Rev. AERE-PEREMNIVS (sic). Saturn, winged and with scythe, lies prostrate. Above him, Minerva, with spear and shield, offers a book to Eternity, who is seated, her feet upon a globe, and pointing to the portals of a temple upon the border of which [tem]PLUM AETERNITATI. Above, a re-entering serpent. Within, in upper right-hand corner, several wreaths. Behind a cliff. EX.M.S.F. 1685."

DIAMETER — 85 mms.

MATERIAL — Bronze.

SHELFMARK — Waller Collection of medals, no. 454.

31.

GOKEL [GOCKELIUS], EBERHARD (1636-1703)

MANUSCRIPT — Latin and German. Unpublished.

TITLE — Compendium oder Kurtzer Bericht von den Brüchen. Wie man dieselbe ohne und durch den Schnitt auch auff magnetische weise curieret und geheglet werden mögen. Verfasset und zusammengetragen von Eberhardo Gockelio, Ulm MED.D. p.t. Stattphysico zu Biengen. Anno reparata salutis 1669.

BOUND WITH — Compendium de lithotomia vescica, das ist Kurtzer Bericht von Blasenstein; and Bericht von den Fontanellen.

PLACE & DATE — Ulm, 1669.

DIMENSIONS — 20.3 x 16.3 cms.

PAGINATION — 36 + 72 + 50 pages.

ILLUSTRATED

BINDING — Contemporary vellum with armorial ex libris.

SHELFMARK — Waller 653 E:2.

HISTORICAL NOTE

Three medical manuscripts illustrated with more than 100 drawings of instruments and surgical operations. Gokel was a physician to the Duke of Württenberg and a member of the *Kaiserlich Leopoldinischen Akademie der Naturforscher*.

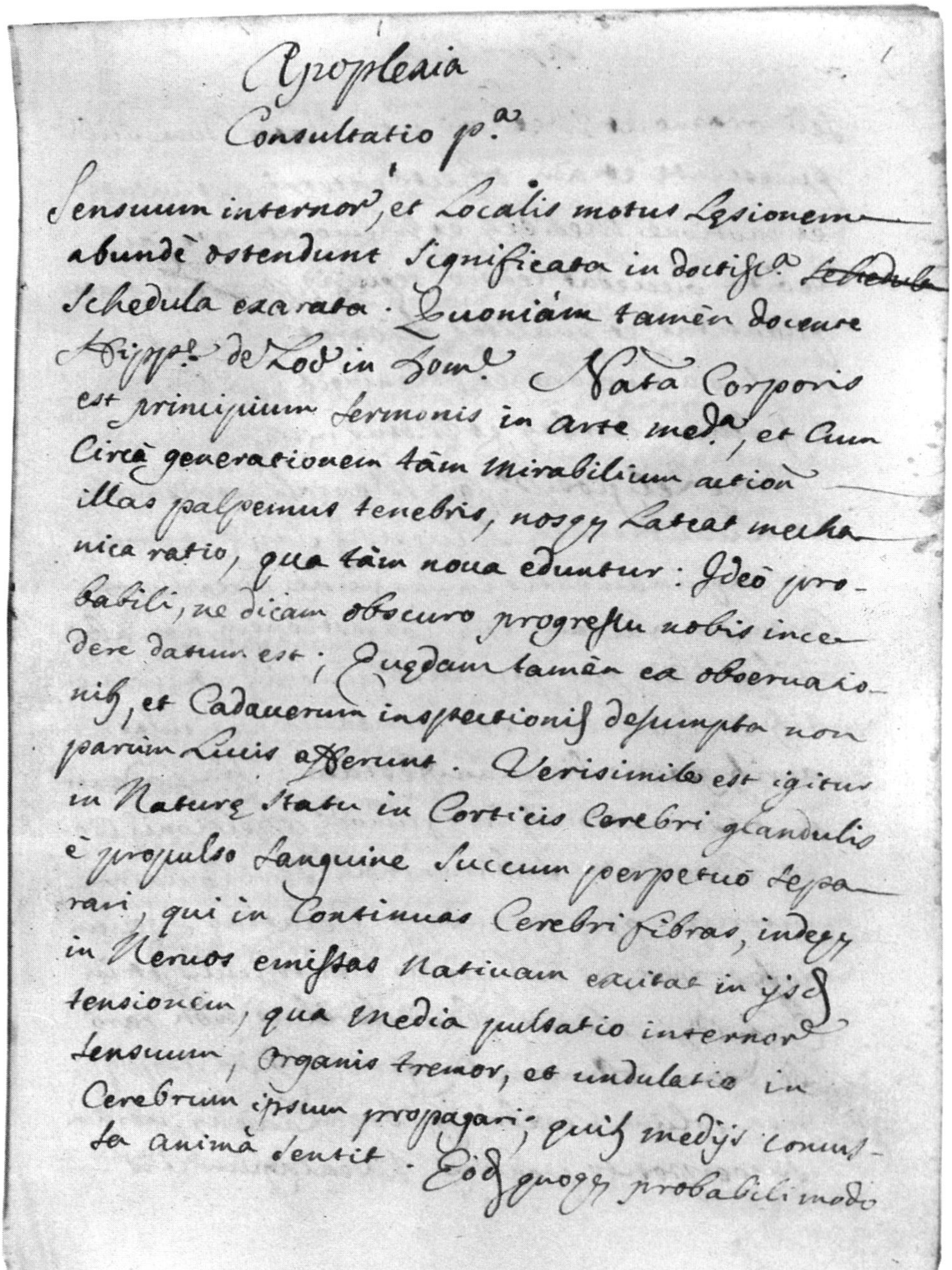

Apoplexia
Consultatio p^a.

Sensuum internor., et Localis motus Lęsionem
abunde ostendunt Significata in doctiss.a ~~Schedula~~
Schedula exarata. Quoniam tamen docente
Hippoc. de Loc. in Hom. Natura Corporis
est principium Sermonis in arte med., et cum
circa generationem tam mirabilium action.
illas palpemus tenebris, nosque lateat mecha-
nica ratio, qua tam noua eduntur: Ideo pro-
babili, ne dicam obscuro progressu nobis ince-
dere datum est; Quędam tamen ex observatio-
nib., et Cadauerum inspectionib. desumpta non
parum Lucis afferunt. Verisimile est igitur
in Naturę statu in Corticis Cerebri glandulis
e propulso Sanguine Succum perpetuo sepa-
rari, qui in Continuas Cerebri fibras, indeque
in Neruos emissas Natiuam excitat in ijsd.
tensionem, qua media pulsatio internor.
Sensuum, Organis tremor, et undulatio in
Cerebrum ipsum propagari; quib. medijs consti-
tuta anima sentit. Eodemque probabili modo

Fig. 33.

32.

MARCHETTI, DOMENICO (1626-1688)

DOCTORAL DIPLOMA — Latin. On vellum. Signed by Domenico Marchetti, Professor of Anatomy at Padua University. The respondent was Calonymos Itala Mantuanus.

PLACE & DATE — Padua, 19.9.1675.

DIMENSIONS — 24 x 17.7 cms.

FOLIATION — 4 leaves.

ILLUMINATED — With broad painted borders in rich colors, including portraits of Aristotle and Hippocrates in contemporary costumes. Large portrait of the doctor and his arms.

BINDING — Original red morocco binding gilt, two seals of the University of Padua and the Venetian Government, in morocco cases gilt.

SHELFMARK — Waller Red morocco box.

33.

[MALPIGHI, MARCELLO (1628-1694)]

MANUSCRIPT — Latin and Italian.

TITLE — Consultationes [medicinales]. In Waller 650 B there are four letters by Malpighi dated 18.1.1674, 29.4 and 27.5 1684 and one undated. In the Waller Collection there is also a letter by Johannes Bohn addressed to Malpighi dated 16.9.1691. All these letters were published by H.B. Adelmann in *The Correspondence of Marcello Malpighi,* 5 vols., (Cornell University Press: Ithaca and London, 1975).

DATE — Different dates (1688-1714).

DIMENSIONS — 19.2 x 14.2 cms.

FOLIATION — 144 leaves.

BINDING — Modern vellum.

SHELFMARK — Waller 653 D:3.

HISTORICAL NOTE

Malpighi, who is primarily known for his mastery of microscopic technique and its successful application to the study of human anatomy, physiology and embriology, was also a renowned practical physician. This manuscript refers to Malpighi's practical medicine and lists 71 medical consultations undertaken

HIPPOCRATES

IN DEI ÆTERNI
NOMINE AMEN

VNIVERSIS, &
singulis presens hoc publicum
Doctoratus Priuilegium visu=
ris lecturis, et audituris NOS
Dominicus de Marchet=
tis Patauinus Phil: et Med:
Doctor, et in Nobilissima Patauina Academ:
ad Lectura Chirurg: Ordinar: in p.º loco pub:
Professor, Inclytique ordinis Dominor Phil:^e
& Medicinæ Doctor, et Professor Auctoritate
Serenis: REIPVBLICÆ VENETÆ
Pro Præses, Salutem in Eo, qui est omnium
vera Salus. Antiqua et preclara Vniuersi=
tas Patauina bonarum literarum mater glo=
riosa, et Artium, ac Scientiarum parens op=
tima, que Clarissimorum Doctor splendore,
Vetustissimor Studiorum auctoritate, et side
reis Doctrine, ac Virtutum omnium monu=
mentis, non modo per Italiam, sed per totam

Fig. 32 a.

Fig. 32 b.

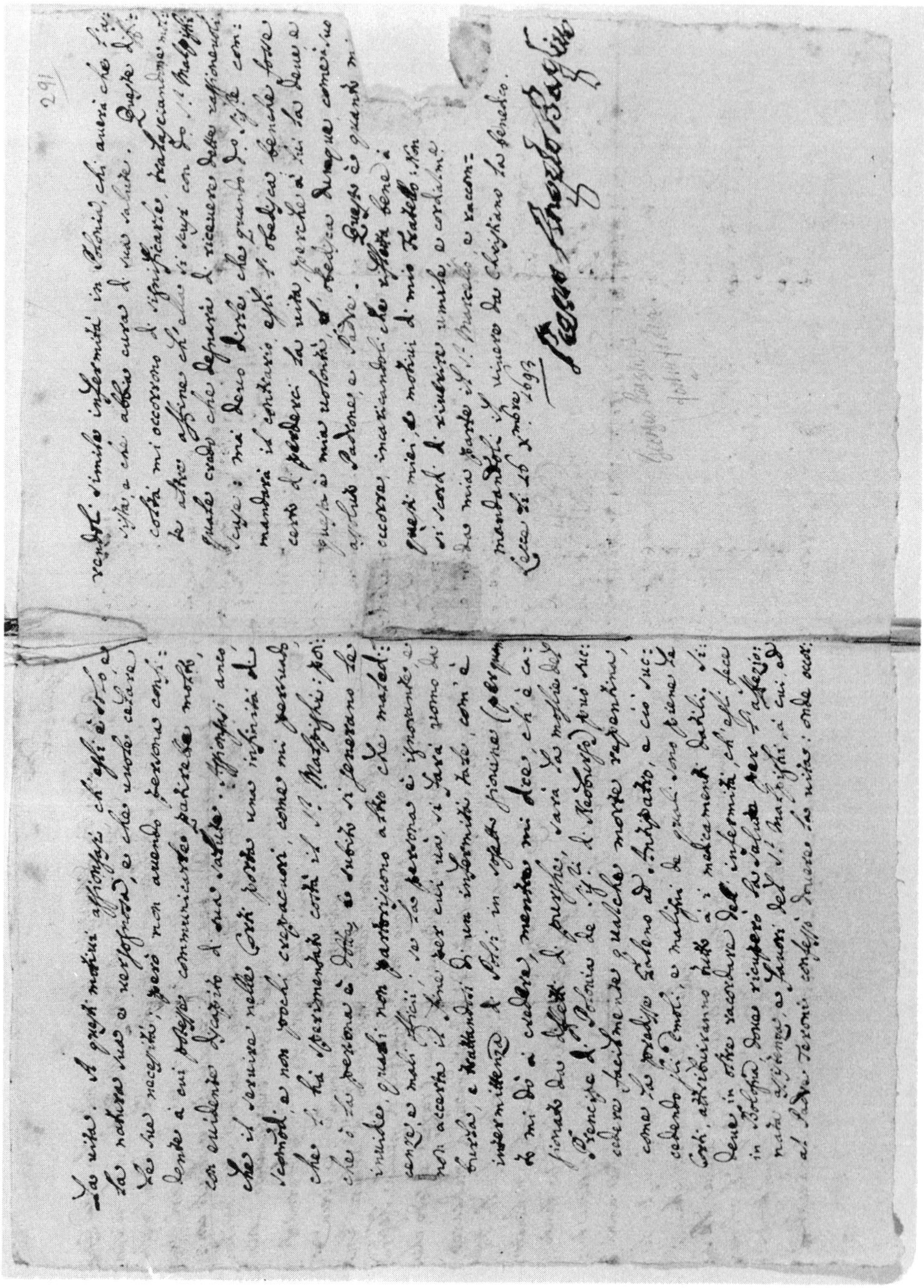

Fig. 34.

by Malpighi during the last period of his life. Although the last ten leaves of the Ms. list a few consultations made by Paganini, Lancisi and Dell'Aqua, most of the Ms. reports those of Malpighi. A first edition of Malpighi *consulti* was not published until 1713 by one of his pupils under the title *Consultationum medicinalium centuria prima,* (Padua, 1713) but it was extremely defective. The most complete collection of Malpighi's *Consultationes* are still in the manuscript versions compiled by his pupils (for a detailed study of the fate of Malpighi's *Consultationes,* see H.B.Adelmann, *Marcello Malpighi and the Evolution of Embriology,* vol. 1, Ithaca-New York, 1966, pp. 665-667). The present Ms. lists among the patients the Duke of Modena, the Archbishop of Milan, the Princess of Vaudemont, Cardinal Pamphili and the King of Poland.

34.

BAGLIVI, GIORGIO (1668-1707)

LETTERS
Collection of 192 letters bound in one volume mostly addressed to Baglivi, written in various languages, mostly Latin and Italian, from different European naturalists. Among the most important are the letters by A. Vallisnieri, B. Ramazzini, A. Magliabechi, W. Musgrave, J.J. Manget, G. Mattei, P. Hotton, G.C. Osterchamp, A. Pacchioni, A.M. Valsalva, P.A. Baglivi, W. Sherard, G. Grandi, L. Tronfietti, J.G. Berger, G. Lanzoni, J.A. Chouet, D. Leclerc. The letters often bear drafts of Baglivi's replies.

DATE
1693-1704.

DIMENSIONS
The dimensions of the volume are 30 x 19.5 cms.

FOLIATION
405 leaves.

SHELFMARK
Waller 653 C:1.

HISTORICAL NOTE

This interesting correspondence, covering the last period of the life ot the Italian physician Giorgio Baglivi, provides valuable and original information on the medical debates at the end of the XVIIth century. In order to understand the importance of this correspondence it is worth mentioning that *Baglivi's Correspondence from the Library of Sir William Osler,* edited by Dorothy M. Schullian (Cornell University Press: Ithaca, 1974) lists 173 letters covering the period 1677-1692. Interestingly, the Waller collection of Baglivi's letters, which are dated to 1693-1707, seems to fill the chronological gap in Osler's collection.

During the International Congress of the History of Medicine held in Zagreb in 1938, Waller described his collection of Baglivi's correspondence. After Waller's death in 1955, the Italian historian of medicine Ladislao Münster planned to publish the correspondence but the project was never carried out and only the letter of Baglivi's father, Pietro Angelo, was published (see L. Münster, "Giorgio Baglivi Mancato Archiatra della Regina di Polonia", in *Rivista di Storia della Medicina,* 1 (1957), pp. 182-194).

35.

[ARZNEYBUCH]

MANUSCRIPT	German.
TITLE	Buch vieler Geheimer und Rarer Experimenten. Bestehende in Bereitung dess von Vielen gesuchten aber von wenige gefundenen Lapides Philosophici. Wie auch anderen chymischen Versuchen benebst magischen medicinischen Arcanen sowohl vor Menschen als Viehe. Dem noch begefügt verschiedene Curiosa Miscellanea. Durch eigener oder anderer guter Freunde Versuche approbiert.
DATE	c. 1700.
DIMENSIONS	32 x 20 cms.
PAGINATION	356 pages.
ILLUSTRATED	
SHELFMARK	Waller 653 D:2.
CONTENTS	

Richly illustrated alchemical manuscript in which a large number (about 600) of prescriptions (both medical and alchemical) are given. The manuscript also provides plenty of advice for the attainement of the philosophers' stone. To this end several allegorical illustrations and symbols adorn the text.

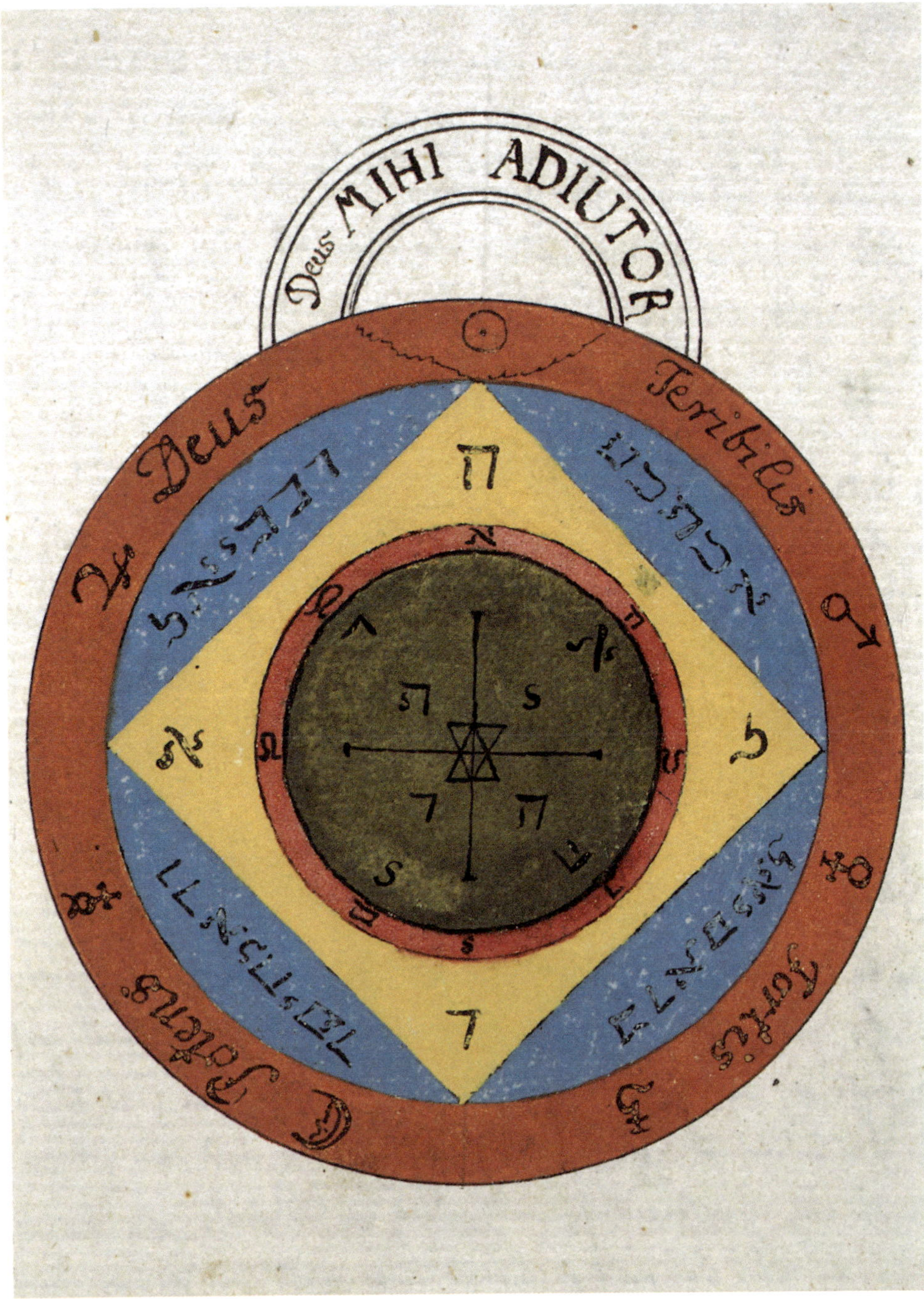

Fig. 35: *Alchemical symbol for the Great Work.*

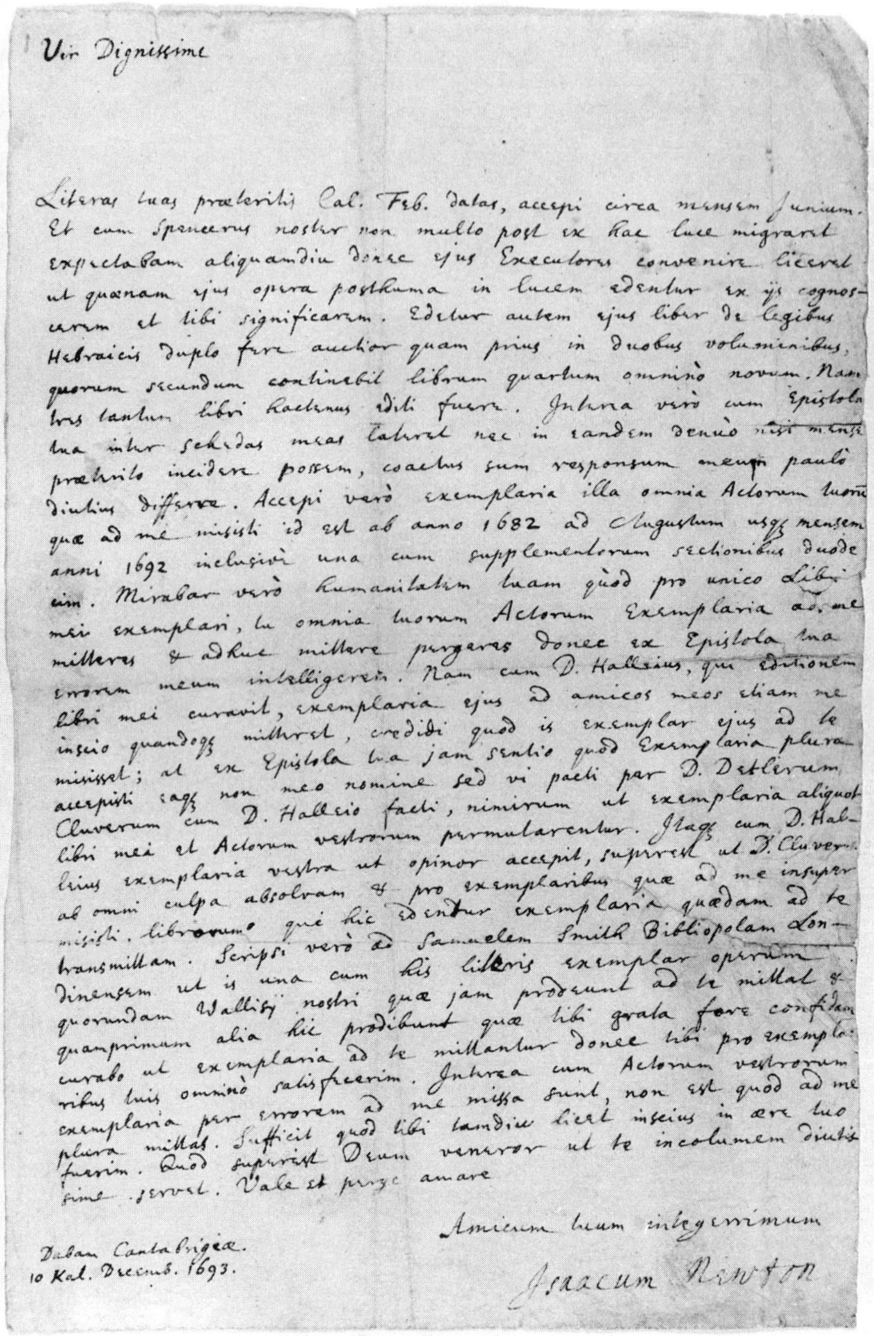
Vir Dignissime

Literas tuas præteritis Cal. Feb. datas, accepi circa mensem Junium. Et cum Spencerus noster non multo post ex hac luce migrarit expectabam aliquamdiu donec ejus Executores convenire liceret ut quænam ejus opera posthuma in lucem edentur ex ijs cognoscerem et tibi significarem. Edetur autem ejus liber de legibus Hebraicis duplo fere auctior quam prius in duobus voluminibus, quorum secundum continebit librum quartum omnino novum. Nam tres tantum libri hactenus editi fuere. Interea vero cum Epistola tua inter schedas meas lateret nec in eandem denuo nisi mense præterito incidere possem, coactus sum responsum meum paulo diutius differre. Accepi vero exemplaria illa omnia Actorum tuorum quæ ad me misisti id est ab anno 1682 ad Augustum usq mensem anni 1692 inclusive una cum supplementorum sectionibus duodecim. Mirabar vero humanitatem tuam quod pro unico Libri mei exemplari, tu omnia tuorum Actorum Exemplaria ad me mitteres & adhuc mittere pergeras donec ex Epistola tua errorem meum intelligerem. Nam cum D. Halleius, qui editionem libri mei curavit, exemplaria ejus ad amicos meos etiam me inscio quandoq mitteret, credidi quod is exemplar ejus ad te misisset; at ex Epistola tua jam sentio quod Exemplaria plura accepisti eaq non meo nomine sed vi pacti per D. Detlerum Cluverum cum D. Halleio facti, nimirum ut exemplaria aliquot libri mei et Actorum vestrorum permutarentur. Itaq cum D. Halleius exemplaria vestra ut opinor accepit, superest ut D. Cluverum ab omni culpa absolvam & pro exemplaribus quæ ad me insuper misisti, librorum qui hic edentur exemplaria quædam ad te transmittam. Scripsi vero ad Samuelem Smith Bibliopolam Londinensem ut is una cum his literis exemplar operum quorundam Wallisij nostri quæ jam prodeunt ad te mittat & quamprimum alia hic prodibunt quæ tibi grata fore confidam curabo ut exemplaria ad te mittantur donec tibi pro exemplaribus tuis omnino satisfecerim. Interea cum Actorum vestrorum exemplaria per errorem ad me missa sunt, non est quod ad me plura mittas. Sufficit quod tibi tamdiu licet inscius in ære tuo fuerim. Quod superest Deum veneror ut te incolumem diutissime servet. Vale et perge amare

Amicum tuum integerrimum

Isaacum Newton

Dabam Cantabrigiæ.
10 Kal. Decemb. 1693.

Fig. 36.

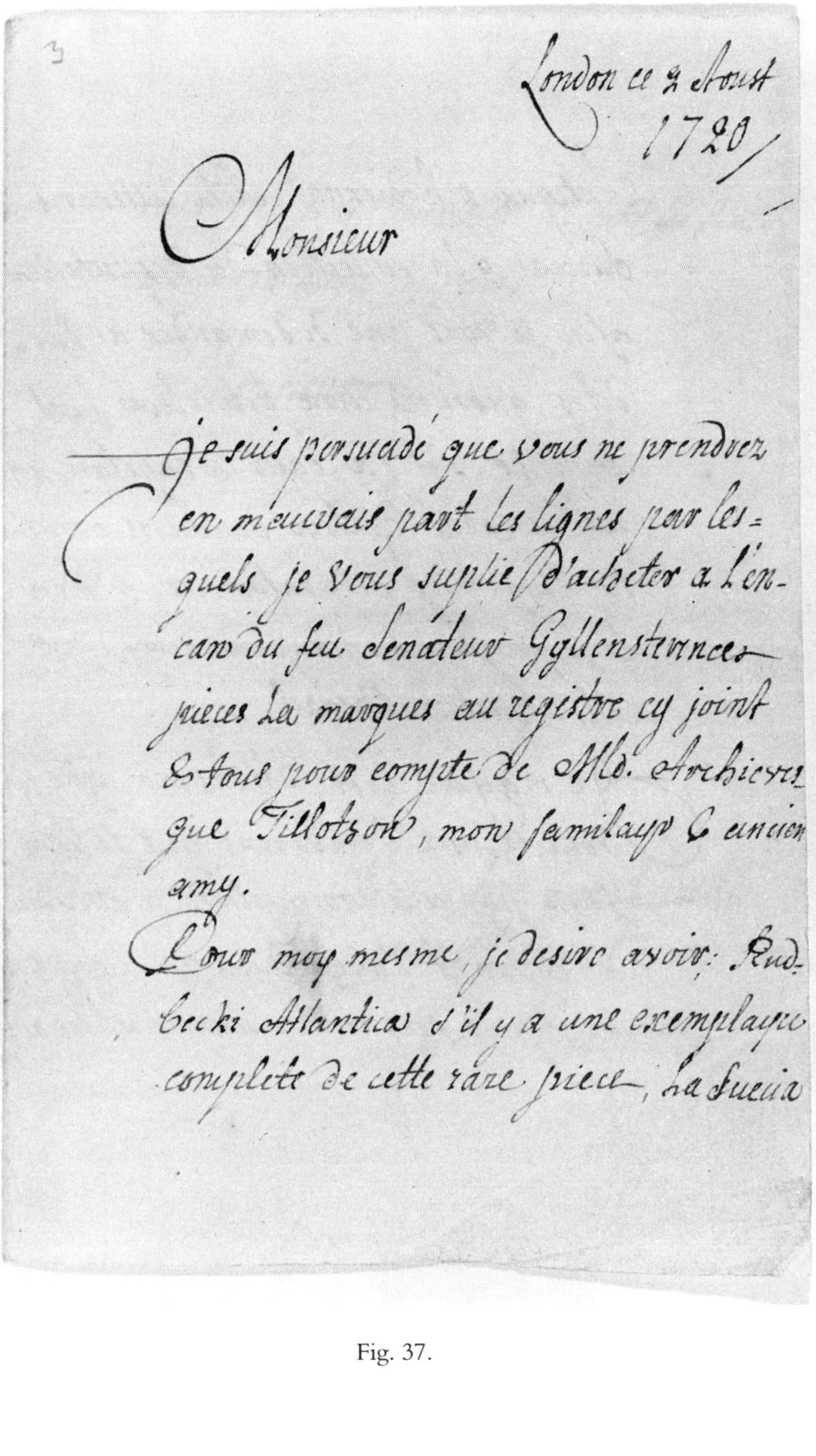

London ce 2 Aoust
1720

Monsieur

Je suis persuadé que vous ne prendrez
en mauvais part les lignes par les=
quels je vous suplie d'acheter a l'en-
can du feu Senateur Gyllenstiernes
pieces la marques au registre cy joint
& tous pour compte de Mld. Archieves-
que Tillotson, mon familayr & ancien
amy.

Pour moy mesme, je desire avoir: Rud-
becki Atlantica s'il y a une exemplayre
complete de cette rare piece, La Suecia

Fig. 37.

36.

NEWTON, SIR ISAAC (1642-1727)

LETTER Latin. Published in *Lychnos,* 1936, pp. 225-229 by Johan Nordström.

PLACE & DATE Cambridge, 10.12.1693.

ADDRESSEE Otto Mencke.

DIMENSIONS 29.8 x 18.8 cms.

FOLIATION 1 leaf.

FOLIATION Waller Delikatesskaps 1.

37.

NEWTON, SIR ISAAC (1642-1727)

LETTER French. Published by A. Rupert Hall in "Further Newton Correspondence", *Notes and Records of the Royal Society of London,* 37 (1982), pp. 28-29. For a detailed description of this file, see the historical note. A fragment of a letter by Newton dated April 22 [1726?] and a receipt signed by Newton on October 3, 1706 are kept in Waller's "Autographes – Medicine", volume no. 40. An autograph by Newton is kept in Christian Ludwig Liptenius' Album amicorum (shelfmark: Y 82b) (see Fig. XV).

PLACE & DATE London, 2.8.1720.

ADDRESSEE [Jombert(?)]

DIMENSIONS 18.6 x 11.8 cms.

FOLIATION 2 leaves.

SHELFMARK Waller 648 B.

TRANSCRIPTION

London ce 2 Aoust
1720

Monsieur

Je suis persuadé que vous ne prendrez en mauvais part les lignes par lesquels je vous suplie d'archeter a l'encan du feu Senateur Gyllensteirnas pieces la marques au registre cy joint & tous pour compte de Mdl. Archievesque Tillotson, mon familayr & ancien amy.

Pour moy mesme, Je desire avoir Rudbecki Atlantica s'il y a une exemplayre complete de cette rare piece. La suecia Antiqua & hodierna, Acta Litteraris Suecia, & la collection des disputations afin je vous prie de demander a Eruditos, quorum curas biblio-

teca sunt commissae, s'il y a dans la capitale, en ms. le codex dialogo de causis erruptae Eloq[uentiae] sive de oratoribus, & d'en le fayre collationer pour moy par Mr. Gustave Benzel. Je m'assure fort, Monsieur, que vous excusez l'embarras que je vous fays par cette commision inattendue an esperant d'avoir dans peu l'occasion ou je puis vous rendre services utiles auctant que je le souhaite [.] Ce sont les sentiments qu'aura toujours
Monsieur
de vostre tres humble
& tous obeissant serviteur
Is. Newton.

HISTORICAL NOTE

Interesting letter in which Newton is asking Jombert[?] to buy some unspecified items at the auction of the estate of the Swedish statesman Nils Gyllenstierna (1648-1720). The English physicist is also looking for a copy of the rare work by Olaus Rudbeck, *Atlantica* (4 vols. 1679-1702), in which Rudbeck tried to demonstrate that Sweden was the original cradle of civilization. This curious work, which was adorned with beautiful woodcuts, soon became an item for bibliophiles. Leibniz owned a copy of it and commented on its contents thoroughly. Rudbeck's *Atlantica* was an extremely rare work and according to Johannes Rudbeck, *Bibliotheca Rudbeckiana,* (Uppsala, 1918), p. 203, only 500 copies of the first volume were printed and most of the copies of the fourth volume were destroyed during the Fire of 1702 which laid in ashes the greater part of Uppsala. Rudbeck's work does not in fact appear in the inventory of Newton's library. In the same letter Newton also asks the Swedish librarian Erik Benzelius (1675-1743) (the name Gustav does not refer to anyone known) to collate a manuscript kept in Stockholm.

The file also contains several letters addressed to Newton (most of which have been either translated into English or described by A.R. Hall) and some documents related to Newton. There are a few letters which seem to be unpublished. The complete list follows the order in which the letters are arranged in the file:

Letter by Pierre Allix in Latin dated 23.10.1713. English abstract in Hall (1982), *op.cit.*, p. 20.

Letter by Nicola d'Aragona written in Italian dated 25.10.1722. Unpublished. Not addressed to Newton?

Letter by Jaques-Antoine Aralud dated 10.10.1722. English abstract in Hall (1982), p. 29.

Letter by Francesco Bianchini in Latin dated April 1714. English abstract in Hall (1982), pp. 20-22. Another undated letter by Bianchini to Newton has been dated to 1713 by Hall, (pp. 18-19) and has been translated into English.

Letter by Johann Petrus Biester in Latin dated 13.8.1725. Translated into English by Hall (1982), pp. 30-31.

Letter by Samuel Crell dated 16.7.1726. Translated into English by Hall (1982), p. 33.

Letter by Thomas Fantet de Lagny in Latin dated 5/16.5.1718. Translated into English by Hall (1982), p. 25.

Undated letter by Georg Petrus Domcke in Latin. English abstract in Hall (1982), p. 34.

Letter by Stefan Engedi in Latin dated 1.8.1720. English abstract in Hall (1982), p. 28.

Letter by François d'Eynaud in Latin dated 6.9.1719. English abstract in Hall (1982), p. 27.

Letter by Guido Grandi in Latin dated 2/13 March 1703. Partial translation in Hall (1982), pp. 15-16.

Letter by Willem Jacob s'Gravesande in French dated 24.6.1718. English translation in Hall (1982), p. 26.

Letter by the Count Ferdinand Ernst de Herbersteins in Latin dated 26.10.1709. Without the name of the addressee but listed among the letters addressed to Newton. English translation in Hall (1982), p. 17.

Letter by Jakob Hermann in Latin dated 11/22.7.1715. Translated into English by Hall (1982), pp. 22-23.

Letter by Antoine de Jussieu in Latin dated 12/23.1.1719. English abstract in Hall (1982), p. 27.

Undated letter by Bartholomaeus Laurentius in Latin. English abstract in Hall (1982), p. 34.

Letter by Johann Georg Liebknecht in Latin dated 30.8/10.9.1723. English abstract in Hall (1982), p. 30.

Letter by Ludovico Antonio Muratori in Latin dated 24.5/4.6.1716. English abstract in Hall (1982), pp. 23-24.

Letter by Petrus van Musschenbroek in Latin dated 12/23.2.1726. English translation in Hall (1982), p. 32.

Letter by Obrecht in French dated 3.10.1722 without the name of the addressee but listed by Waller with the letters addressed to Newton. Unpublished.

Letter by Ludovico a Ripa in Latin dated late 1718. English abstract in Hall (1982), p. 27.
Letter by Albert de Sallengre in Latin dated 27.10/7.11.1718. English translation in Hall (1982), p. 26.
Four letters by Johann Jakob Scheuchzer in Latin dated 3/14.4.1706, 15/26.10.1709, 17/28.10.1709, 22.10/2.11.1709. English abstracts in Hall (1982), pp. 16-17.
Letter by Francesco Torti in Latin dated 20/31.7.1716. English abstract in Hall (1982), p. 24.
Letter by Nicolò Troni in French dated 18.3.1718. Abstract in Hall (1982), p. 25.
Letter by the Marquis Giovanni Giuseppe Orsi in Latin dated 25.4/6.5.1717. English abstract in Hall (1982), p. 24.
Letter by Giuseppe Velletta in Latin dated 2/13.2.1712. English abstract in Hall (1982), pp. 17-18.
Letter by Levin Vincent in French dated 30.8.1726. English abstract in Hall (1982), pp. 33-34.
Two letters by Burchard de Volder in Latin dated 12/24.11.1684 and 4/14.8.1687. English abstracts in Hall (1982), pp. 11-12.
Letter by Daniel Waeywel in Dutch dated 1.2.1726. English abstract in Hall (1982), pp. 31-32.
Letter by Heinrich Wetstein in Latin dated 24.11.1722. English translation in Hall (1982), pp. 29-30.

The file also contains:
Latin manuscript of eight leaves (in Newton's hand?) in which some theological questions are discussed.
Three letters written in French by an anonymous author in which the name and works of Newton are mentioned.
An unsigned letter in Latin dated 26.4.1671.

38.

[NEWTON, SIR ISAAC (1642-1727)]

COMMEMORATIVE MEDAL

described in Hippolyte Kluyskens, *Des Hommes Célèbres dans les Sciences et les Arts, et des médailles qui consacrent leur souvenir*, (Gend, 1859), vol. 2, p. 248.

Ob. ISAACUS. NEWTONIUS. Bust to right. Made by J. Dassier

Fig. 38.

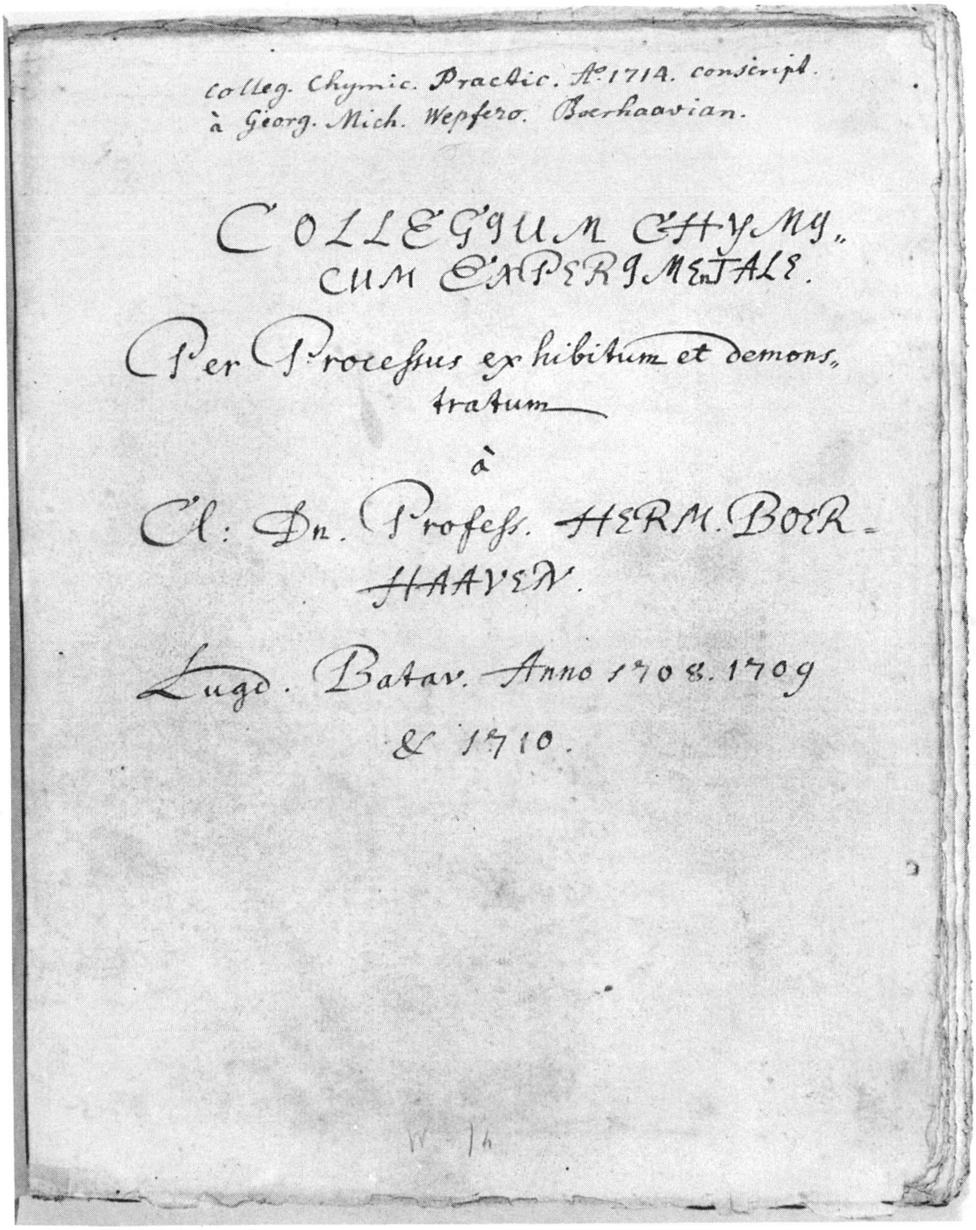

Colleg. Chymic. Practic. Aº 1714. conscript.
à Georg. Mich. Wepfero. Boerhaavian.

COLLEGIUM CHYMY-
CUM EXPERIMENTALE.

Per Processus exhibitum et demons-
tratum

à

Cl: Dn. Profess. HERM. BOER-
HAAVEN.

Lugd. Batav. Anno 1708. 1709
& 1710.

Fig. 39.

possibly before 1739.
Rev. NAT. 1642. M. 1726 [sic] and a mausoleum.

DIAMETER 42 mms.
MATERIAL Bronze.
SHELFMARK Waller Collection of medals, no. 372.

39.

[BOERHAAVE, HERMANN (1668-1738)]

MANUSCRIPT Latin. Unpublished.

TITLE Collegium Chymicum experimentale. Per processus exhibitum et demonstratum a Cl: Dn. Profess. Herm. Boerhaaven. In the same box (653 D:3) there is a manuscript version of Pars Secunda Pracxios Medico Chirurgia ex Praelextionibus, celebr. Dom. Professor Boerhave collecta ex lingua latina in maternam Svetica a Olos Joh. Acrell [dated 1757-1758]. In Waller 649 F:1 there are eight autograph letters dating from 1714-1733.

PLACE & DATE Lugduni Batavorum, 1708, 1709, 1710.

DIMENSIONS 21.1 x 16.5 cms.

PAGINATION 6 + 434 pages.

SHELFMARK Waller 653 D:3.

HISTORICAL NOTE

This manuscript is a collection of notes taken by the Dutch physician Georg Michael Wepfer during the chemical lectures held by Boerhaave in Leiden in the years 1708, 1709 and 1710. The manuscript was written in 1714 (see p. 437) and it reports an extremely rich and detailed collection of chemical experiments. The manuscript differs from Boerhaave's *Institutiones et Experimenta chemiae,* a series of lectures published in 1724 by his students but bitterly criticized by Boerhaave himself. The course in chemistry, however, follows the traditional division into three natural kingdoms. The first 221 pages deal with the plant kingdom and provide a detailed description of chemical operations related to it, i.e. distillation, fermentation, extraction, etc. The second part deals with a series of *experimenta in Animalibus* (pp. 222-299); the last part deals with the *fossilium historia et Analysis chemica* (the history of minerals and chemical analysis) (pp. 300-434).

40.

[ASTRUC, JEAN (1684-1766)]

MANUSCRIPT French.

TITLE Maladies des femmes expliquées par Mr. Astruc. Professeur au College Roial de Paris. 1739 et 1740.

PLACE & DATE [Paris], 1739 and 1740.

DIMENSIONS 18.2 x 12.5 cms.

PAGINATION (4) + 426 + (1) + 195 pages.

BINDING Contemporary leather binding.

SHELFMARK Waller 653 E:2.

HISTORICAL NOTE

This Ms. is a collection of notes presumably taken by a pupil of Astruc during a course held at the *Collège Royale de Médicine* in Paris. Astruc published his gynaecological lectures between 1761 and 1765 under the title *Traité sur les maladies des femmes* (Paris, 6 vols.). On page 14 of the first volume of the printed version he declared: "J'ai enseigné depuis longtemps au Collège Royal de France ce Traité que je publie aujourd'hui et je l'ai enseigné à plusieurs reprises". In fact Astruc is known for having been a successful teacher of medicine.

41.

NEEDHAM, JEAN TURBERVILLE, (1713-1781)

LETTER French. The file contains two more letters.

PLACE & DATE Geneva, 3.12.1765.

ADDRESSEE Sabatié.

DIMENSIONS 31 x 20.6 cms.

FOLIATION 2 leaves.

SHELFMARK Waller 648 B:2.

HISTORICAL NOTE

Brought up in a religious and pious environment, Needham (an English Catholic) associated his research on the theory of reproduction and his use of the microscope in biology with his deep religious faith. Needham's "scientific interests were largely motivated by a desire to defend religion in an age when biological questions had serious theological and philosophical meanings for many"(*DSB*, article "Needham"). On more than one

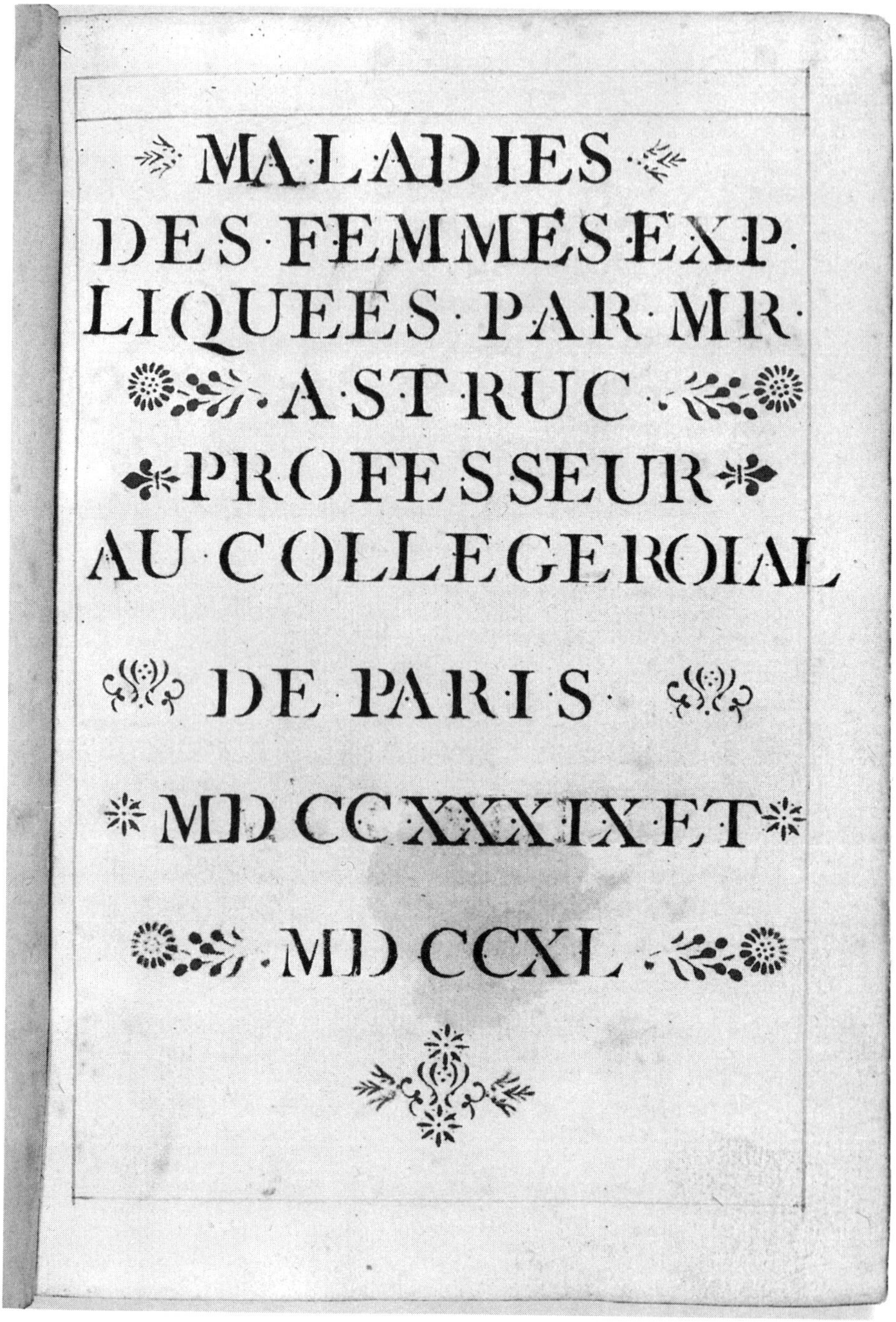
MALADIES
DES FEMMES EXP
LIQUEES PAR MR
ASTRUC
PROFESSEUR
AU COLLEGE ROIAL
DE PARIS
MD CCXXXIX ET
MD CCXL

Fig. 40.

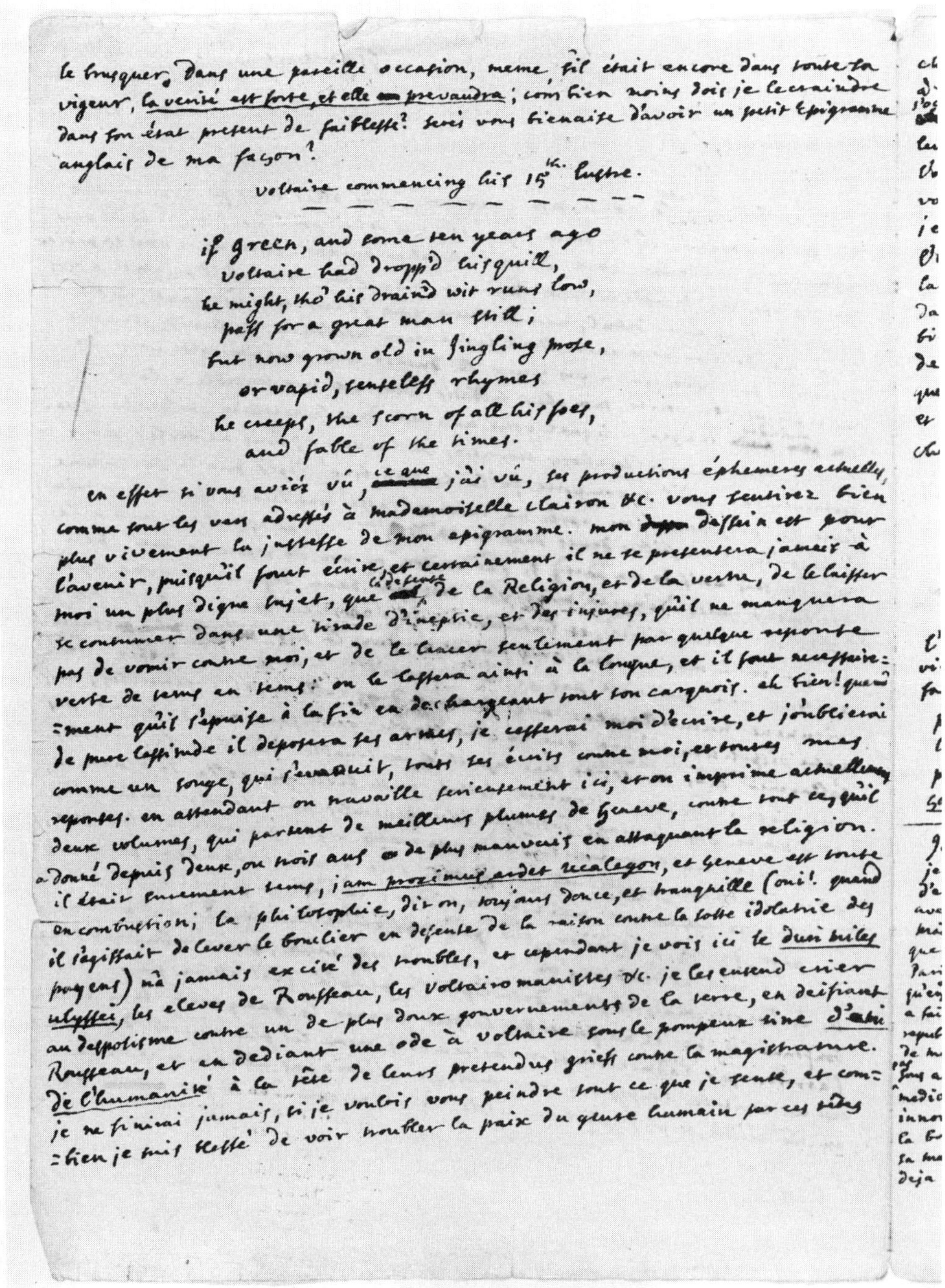
le brusquer, dans une pareille occasion, meme, s'il était encore dans toute sa
vigeur, la verité est forte, et elle prevaudra; com bien moins dois je le craindre
dans son état present de faiblesse? serés vous bienaise d'avoir un petit Epigramme
anglais de ma façon?

voltaire commencing his 15th. lustre.

if green, and some ten years ago
voltaire had dropp'd his quill,
he might, tho' his drain'd wit runs low,
pass for a great man still;
but now grown old in jingling prose,
or vapid, senseless rhymes
he creeps, the scorn of all his foes,
and fable of the times.

en effet si vous aviés vú, ce que j'ai vú, ses productions éphemeres actuelles,
comme tous les vers adressés à mademoiselle clairon &c. vous sentirez bien
plus vivement la justesse de mon epigramme. mon dessein est pour
l'avenir, puisqu'il faut écrire, et certainement il ne se presentera jamais à
moi un plus digne sujet, que la defense de la Religion, et de la vertu, de le laisser
se continuer dans une tirade d'ineptie, et des injures, qu'il ne manquera
pas de vomir contre moi, et de le lancer seulement par quelque reponse
verte de tems en tems: on le lassera ainsi à la longue, et il faut necessaire=
=ment qu'il s'epuise à la fin en dechargeant tout son carquois. eh bien! quand
de pure lassitude il deposera ses armes, je cesserai moi d'écrire, et j'oublierai
comme un songe, qui s'évanouit, touts ses écrits contre moi, et toutes mes
reponses. en attendant on travaille serieusement ici, et on imprime actuellement
deux volumes, qui partent de meilleures plumes de Geneve, contre tout ce qu'il
a donné depuis deux, ou trois ans de plus mauvais en attaquant la religion.
il était vraiment tems, jam proximus ardet ucalegon, et Geneve est toute
en combustion; la philosophie, dit on, toujours douce, et tranquille (oui! quand
il s'agissait de lever le bouclier en defense de la raison contre la sotte idolatrie des
payens) n'a jamais excité des troubles, et cependant je vois ici le duri biles
ulysses, les eleves de Rousseau, les voltairomanistes &c. je les entend crier
au despotisme contre un de plus doux gouvernements de la terre, en deifiant
Rousseau, et en dediant une ode à voltaire sous le pompeux titre d'ami
de l'humanité à la tête de leurs pretendus griefs contre la magistrature.
je ne finirai jamais, si je voulois vous peindre tout ce que je sens, et com=
=bien je suis blessé de voir troubler la paix du genre humain par ces têtes

Fig. 41.

ANDREAS CELSIUS ASTRON. PROF. UPSAL.
REGIA ACADEMIA SCIENTIARVM.

Fig. 42.

occasion Needham defended the authority of the Holy Scriptures from the attacks made by Voltaire. This letter is an interesting testimony to Needham's beliefs. In it the doctrines of Rousseau and Voltaire are heavily criticized and regarded as a threat to religion and to the Government of Geneva. Of particular interest is the following epigram against Voltaire which was composed by Needham in English:

Voltaire commencing his 15th lustre
– – – – – – – – – – – – – – – – – – – –

if green, and some ten years ago
Voltaire had dropp'd his quill,
he might, tho' his drain'd wit runs low,
pass for a great man still;
but now grown old in tingling prose,
or vapid, senseless rhymes
he creeps, the scorn of all his foes,
and fable of the times.

42.

[CELSIUS, ANDERS (1701-1744)]

COMMEMORATIVE MEDAL	Described by Kluyskens vol. 1, p. 192 (see item No. 38).
	Ob. ANDREAS CELSIUS ASTRON. PROF. UPSAL. Bust to right. Below D.F. (Daniel Fehrmann).
	Rev. The Polar Star surrounded by three crowns and the arms of the Royal Academy of Sciences of Stockholm REGIA ACADEMIA SCIENTIARUM.
DIAMETER	30 mms.
MATERIAL	Silver.
SHELFMARK	Waller Collection of medals, no. 1004.

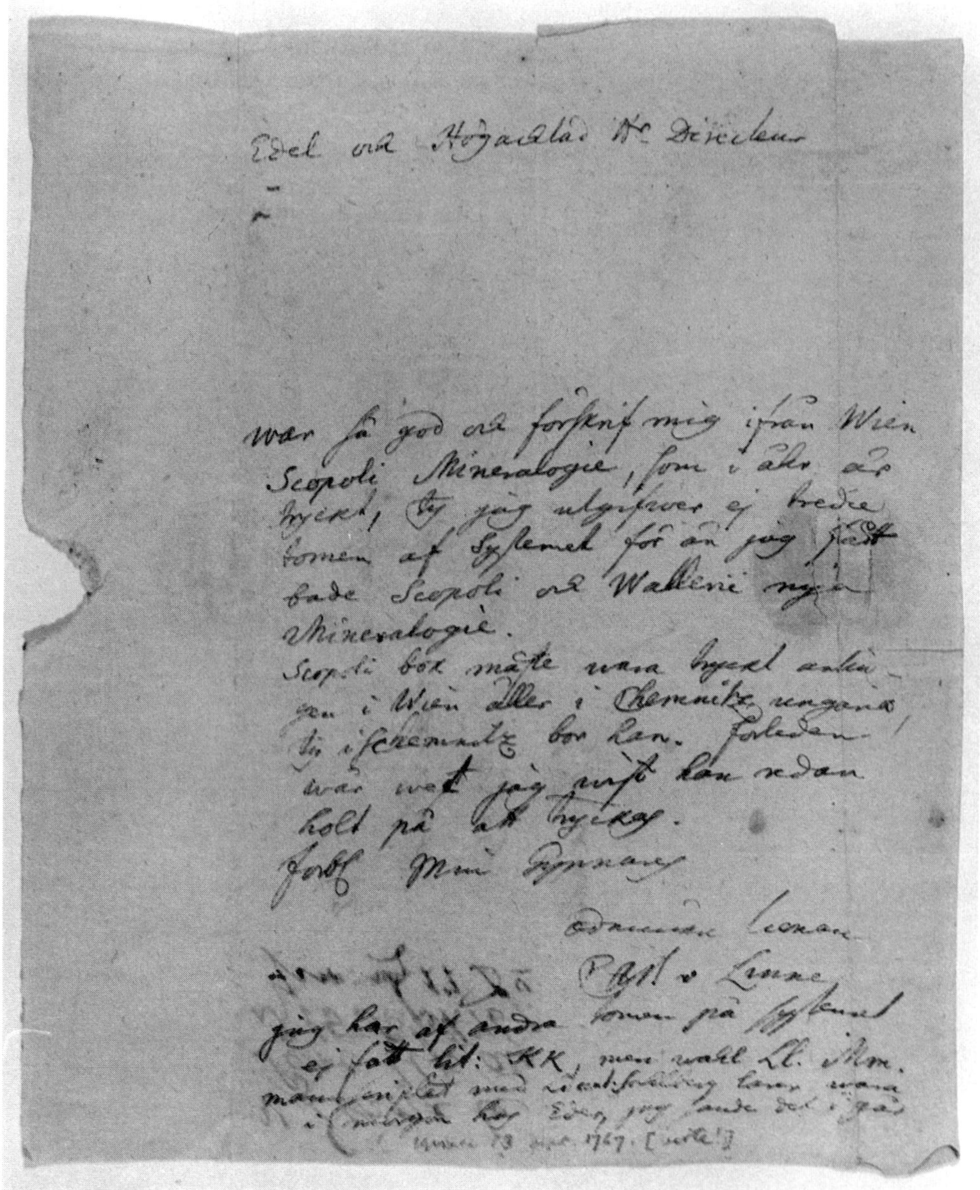

Edel och Högaktad Hr Directeur

war så god och förskrif mig ifrån Wien
Scopoli Mineralogie, som i åhr är
tryckt, ty jag utgifwer ej tredie
tomen af Systemet förän jag fådt
både Scopoli och Walleri nya
Mineralogie.
Scopoli bok måste wara tryckt antin-
gen i Wien eller i Chemnitz ungarns
ty i Chemnitz bor han. Förleden
wår wet jag wist han redan
höllt på att tryckas.
Förbl. Min Gynnares
ödmiukaste tienare
Carl v Linné
jag har af andra tomen på Systemet
ej fått lit: KK, men wäl Ll. Mm.
[illegible]
i morgon kan Eder, jag sände det i går
Upsala 13 apr. 1767. [illegible]

Fig. 43.

43.

LINNAEUS, CAROLUS (1707-1778)

LETTER
Swedish. Unpublished.
The file also contains three more letters and thirteen autographs (some of which are related to the courses delivered by Linnaeus at Uppsala University); 1 letter by H.J.N. Crantz to Haller (1764); 5 letters by J. Dryander 3, (1788), (1794), and (1805); 1 letter by P. Camper to J. Smith and Smith's reply concerning membership of the Linnean Society of London both dating to 1788; 1 letter by Sam. Linnaeus (dated 9.4.1778); 1 letter by Sara E. Morea (dated 1777); 1 diploma of the *Société Linnéenne de Paris* (see next item).

PLACE & DATE
[Uppsala, 13.9.1767.]

ADDRESSEE
Lars Salvius (publisher in Stockholm).

DIMENSIONS
20.2 x 16 cms.

FOLIATION
2 leaves.

SHELFMARK
Waller 652 D:2.

TRANSCRIPTION

Edel och Högacktad Hr Directeuren

war så god och förskrif mig ifrån Wien Scopoli Mineralogie, som i åhr är tryckt, ty jag utgifwer ej tredie tomen af Systemet för än jag fått både Scopoli och Wallerii nya Mineralogie.
Scopoli bok måste wara tryckt antingen i Wien äller i Chemnitz, ungaria, ty i Schemnitz bor han. Förleden wår wet jag wist han redan holt på att tryckas
förbl. Min Gynnare
ödmiuke tienare
Carl v. Linné
Jag har af andra tomen på systemet ei fått hit: KK, men wahl Ll. Men manuscriptet med Licent. Sahlberg lärer wara i morgon hos Eder, jag sände det i går.

Directeuren
Edel och Högacktad
Hr Lars Salvius
Stockholm

HISTORICAL NOTE

Linnaeus, who is preparing the publication of the third volume of a new edition of his *Systema Naturae* (devoted to the mineral kingdom), asks his publisher Lars Salvius to get him the mineralogical works of Giovanni Antonio Scopoli and of Johann

TORBERNUS BERGMAN CHEM·PROF·UPS
LUCTUS TU QUOQUE CAUSSA MEI
ACAD·SCIENT·
HOLM·1785

Fig. 45.

Gottschalk Wallerius, both followers of his taxonomical method. This letter confirms that most of Linnaeus' classification of minerals relied on the works of other naturalists.

44.

[LINNAEUS, CAROLUS (1707-1778)]

DIPLOMA "Diplome de la Société Linnéenne de Paris" belonging to the French naturalist Sebastien Léman. Signed by Bernard-Germain-Étienne, comte de Lacépède.

PLACE & DATE Paris, 10.1.1822.

DIMENSIONS 37.2 x 51 cms.

SHELFMARK Waller 652 D:2.

HISTORICAL NOTE

The *Société Linnéenne de Paris* was the first scientific society to bear the name of the Swedish naturalist. The Society was founded in December 1787 by the French naturalist Pierre Auguste Broussonet who intended to bring about the diffusion of Linnaeus' sexual system at an institutional level. After some problems the *Société* was revived by Thiébaut de Bensaud in 1820 and the Presidency was awarded to Lacépède, a member since 1787. Among the most illustrious members of this second phase were Cuvier, Geoffroy Saint-Hilaire and A. Von Humboldt. In the Diploma are shown the portraits of Tournefort (to the left) and Linnaeus (to the right) with their respective botanical emblems.

45.

[BERGMAN, TORBERN (1735-1784)]

COMMEMORATIVE MEDAL described by Storer n. 298 (see item No. 9).

"Ob. TORBERNUS BERGMAN CHEM. PROF. UPS. Decorated bust to right. Below G.L.

Rev. LUCTUS TU QUOQUE CAUSSA MEI. Between cypresses an urn bearing D. 8. IUL. 1784 beneath crossed oak and palm. EX ACAD. SCIENT. HOLM. 1785."

DIAMETER 33 mms.

MATERIAL Silver.

SHELFMARK Waller Collection of medals, no. 2014.

Théophraste Malpighi Hales
LINNÉ
Bergmann Werner Dolomieu
FAMAM EXTENDERE FACTIS.
Hippocrate Haller Boërhaave
Lavoisier Fourcroy Parmentier
Fabricius Swammerdam Réaumur
Aristote Pline Buffon
TOURNEFORT
Columella Olivier de Serres Crescenzio
La Société Linnéenne de Paris,
Le Président
Le Premier Vice-Président
Le Second Vice-Président
Le Secrétaire perpétuel

Fig. 44.

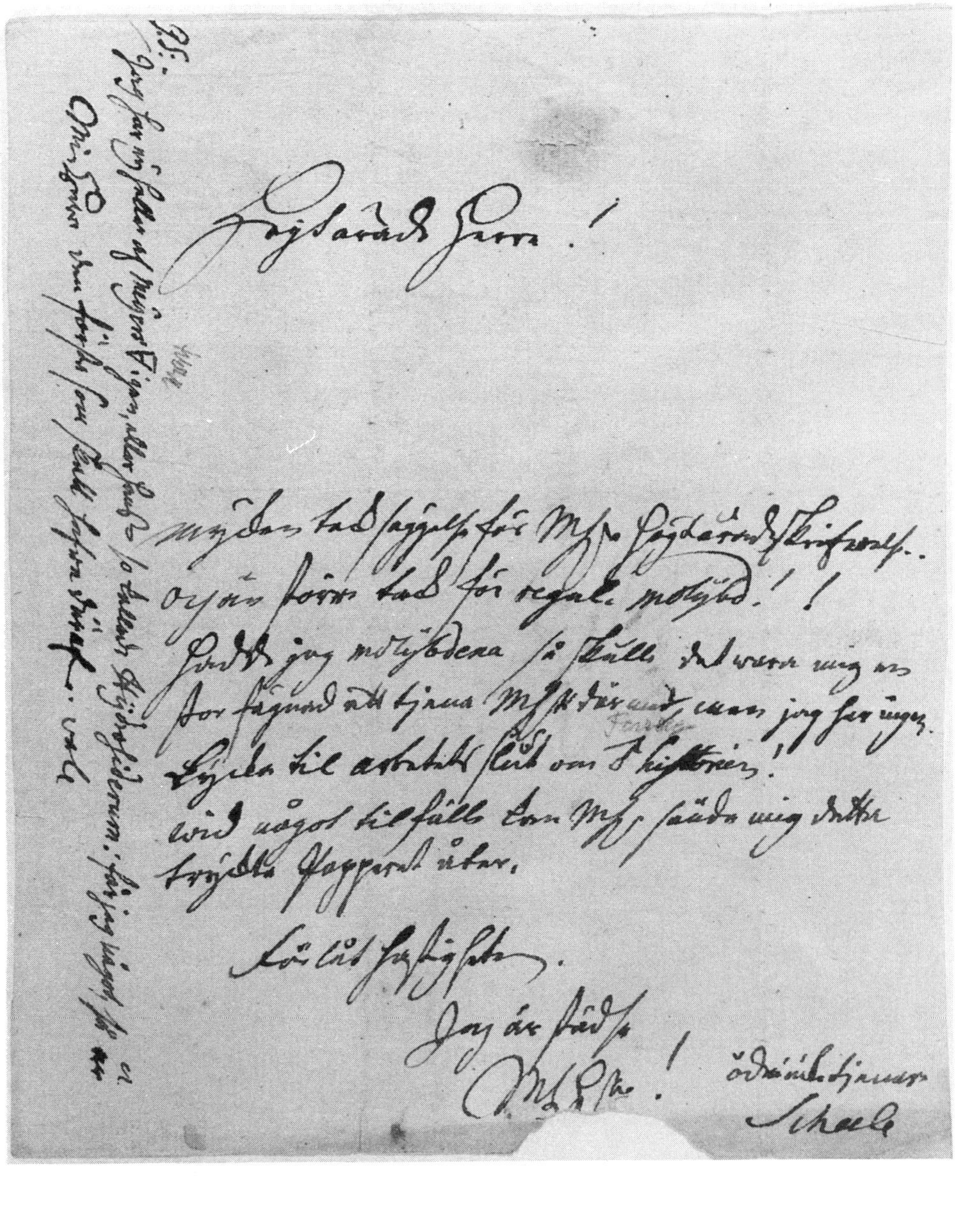

Fig. 46.

46.

SCHEELE, CARL WILHELM (1742-1786)

LETTER	Swedish. Unpublished.
PLACE & DATE	[Köping, after 1781.]
ADDRESSEE	Monsieur de Hjelm [Peter Jacob Hjelm 1746-1813], Stockholm.
DIMENSIONS	20 x 16 cms.
FOLIATION	1 leaf.
SHELFMARK	Waller Delikatesskaps 1.

TRANSCRIPTION

Högtärade Herre!
Mycken tack säggelse för [min herres] högtärade skrifwelse och är stora tack för regali molÿbd!! [molybdena]
Hadde jag molybdena så skulle det wara mig en stor fägnad att tjäna [min herre] därmed men jag har ingen. Lycka til arbetets slut om ♂ [iron] historien!

Wid något tilfälle kan [min herre] sända mig detta tryckta papperet åter.
förlåt hastigheten

Jag är städse
[Min herres]! ödmiuke tjenare
Scheele

P.S. Jag har eij heller af Meyers ∀ [terra pinguis] igen eller hans so kallade hÿdrosiderum : får jag något så er min Herre den förste som skall hafwa däraf. Vale.

HISTORICAL NOTE

Letter in which Scheele thanks his colleague, the chemist P.J. Hjelm, for the specimen of metallic molybdenum. Hjelm obtained metallic molybdenum in 1781 but it was Scheele who in 1778 distinguished between the two minerals, molybdenite and graphite, and gave the inspiration to Hjelm's discovery. It seems likely that this was the first time Scheele was able to see a specimen of metallic molybdenum.

47.

[HUNTER, WILLIAM (1718-1783)]

MANUSCRIPT English.

TITLE Lectures Anatomical and Chirurgical By William Hunter Physician Extraordinary to Her Majesty, Professor of Anatomy to the Royal Academy and Fellow of the Royal and Antiquarian Societies. 1775.

DATE 1775.

DIMENSIONS 24 x 18.4 cms.

FOLIATION 206 leaves.

BINDING Contemporary half calf.

SHELFMARK Waller 653 D:2.

HISTORICAL NOTE

William Hunter was one of the most important anatomists of the second half of the XVIIIth century. The whole Ms. is devoted to pregnancy, abortions, delivery, complications related to pregnancy and other gynaecological diseases. One year before this Ms. was compiled, Hunter published *The Anatomy of the Human Gravid Uterus*, (Birmingham, 1774), (*Bibliotheca Walleriana*, no. 5004 Ligg.Fol.), a superb work in large folio illustrated with 34 engravings.

48.

HUNTER, WILLIAM (1718-1783)

ENGRAVING OF WILLIAM HARVEY

with the signatures of William Hunter and William Hewson. The caption of the engraving is "These are to certify that Mr John Baker, surgeon hath diligently attended a Course of our Anatomical and Chirurgical Lectures. London 27 Jan. 1772". The file also contains two letters by Hunter.

DIMENSIONS 30 x 19 cms.

SHELFMARK Waller 648 C:1.

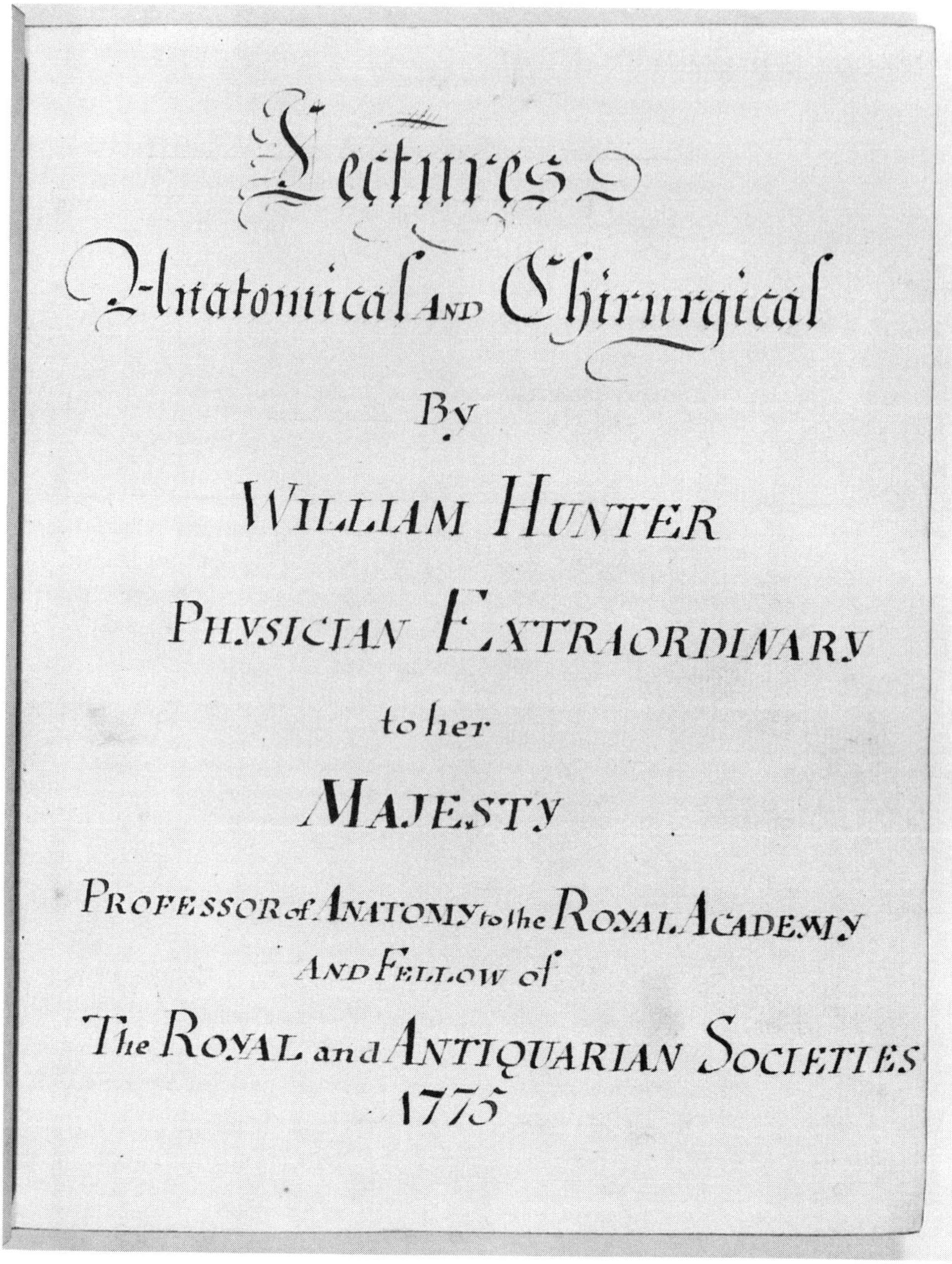

Lectures
Anatomical and Chirurgical
By
WILLIAM HUNTER
PHYSICIAN EXTRAORDINARY
to her
MAJESTY
PROFESSOR of ANATOMY to the ROYAL ACADEMY
AND FELLOW of
The ROYAL and ANTIQUARIAN SOCIETIES
1775

Fig. 47.

WILLIAM
HARVEY
M.D.
These are to Certify that
Mr. John Baker, Surgeon
hath diligently attended a Course
of our Anatomical and Chirurgical
LECTURES. William Hunter.
London 27 Jan. 1772. William Hewson

Fig. 48.

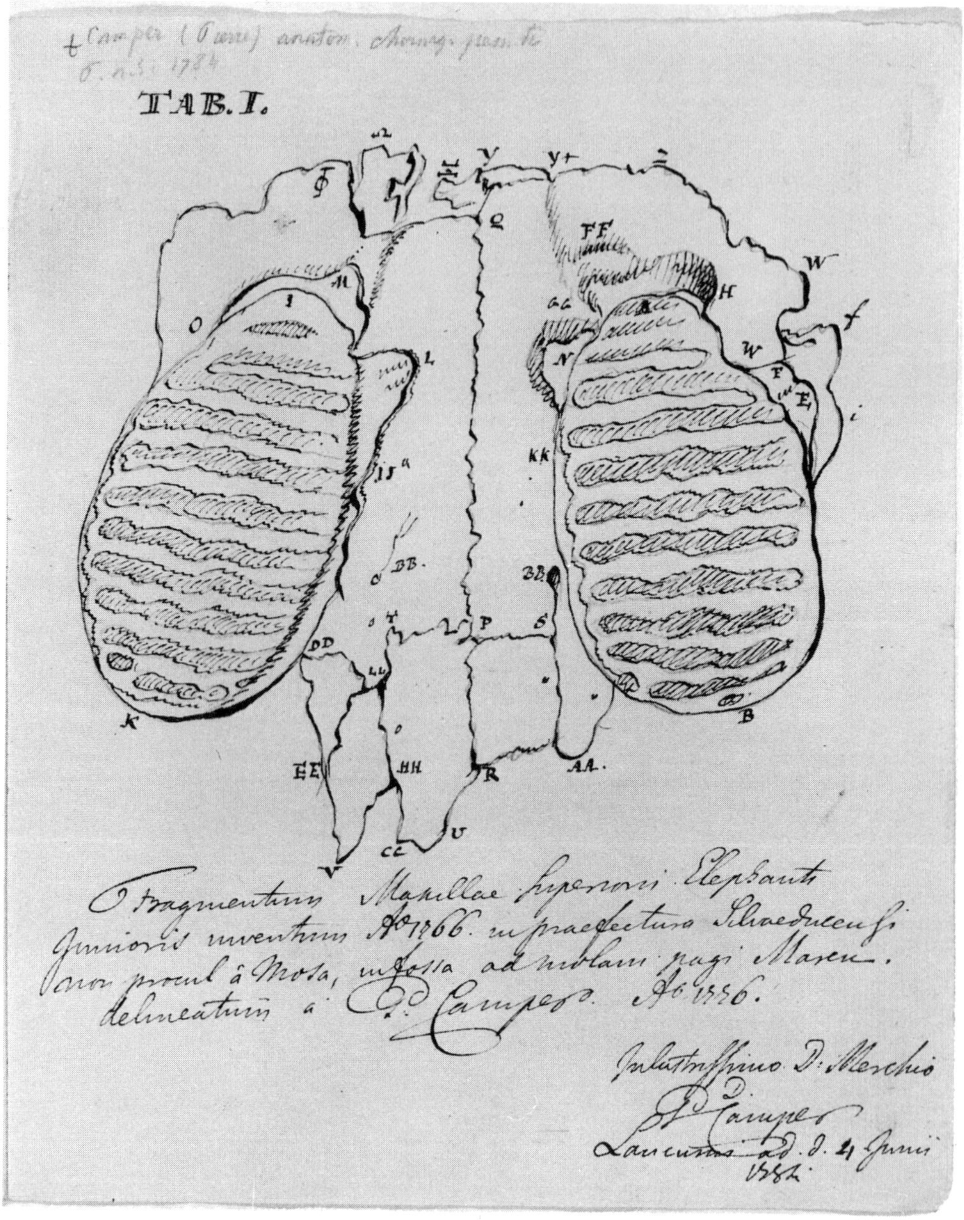

Fig. 49.

49.

CAMPER, PETER (1722-1789)

DRAWING OF A FRAGMENT OF THE JAW OF AN ELEPHANT. In the file there are also six autograph letters. Another letter by Camper can be found in the file Linneana shelfmark: Waller 652 D:2. An undated manuscript in Dutch dealing with surgical operations [Chirurgicale Operatien], possibly in Camper's hand, is kept in Waller 653 D:1.

DATE 4.6.1784.

DIMENSIONS 27.1 x 21 cms.

FOLIATION 2 leaves.

SHELFMARK Waller 649 F:1.

HISTORICAL NOTE

Professor of Anatomy and Surgery at the Athenaeum Illustre in Amsterdam from 1755, Camper conducted pioneering research in comparative anatomy. As reported in *DSB* (article "Camper") "Camper has a great talent for illustrating, as is shown in many of his publications [...] He discussed the relations between anatomy and the arts of drawing and painting in a two-volume treatise that was translated by Cogan five years after his death". This work was *The Works of the late Professor Camper, or the connexion between the Science of Anatomy and the Art of Drawing, Painting, Statuary,* (London, 1794).

50.

ADANSON, MICHEL (1727-1806)

SPECIMEN OF DRIED FISH captured off the island of Tenerife glued on a paper leaf.

DIMENSIONS 20.9 x 16.6 cms.

SHELFMARK Waller 648 E:1. There is an autograph letter by Adanson in the same file.

HISTORICAL NOTE

During the period 1748-1754 the famous botanist and distinguished member of the Royal Academy of Sciences in Paris, Michel Adanson made several journeys in Africa, namely Senegal, and he described its flora and fauna. It is likely that this fish was collected by Adanson off Tenerife during his trip to Africa. During the XVIIIth century various techniques were developed

Fig. 50.

to preserve natural specimens from destruction. The increasing popularity of curiosity and natural cabinets and the mutual exchange of natural specimens between the naturalists of Europe, made the refinement of method of conservation necessary. The high quality of the method of conservation allowed Adanson, as well as most Europen naturalists, to enrich their works with realistic and extremely effective illustrations.

51.

[BLACK, JOSEPH (1728—1799)]

MANUSCRIPT English. Unpublished.
TITLE Lectures upon Chemistry by Dr. Black in 3 volumes.
PLACE & DATE Edinburgh, 1768-1769.
DIMENSIONS 23.2 x 18.4 cms.
FOLIATION 3 vols. 225, 188, 225 leaves.
ILLUSTRATED
SHELFMARK Waller 653 D:3.
HISTORICAL NOTE

An early version of Black's famous lectures on chemistry. Joseph Black, who was one of the most prominent chemists of the XVIIIth century, never published his course of lectures. A posthumous version of his lectures was published by one of his pupils, Robinson, in 1803. There are several Ms. copies of Black's lectures on chemistry, especially in British libraries. However, it is rare to find a copy so rich in details and so clearly written. Furthermore, the date of Waller's copy is exceptionally early and goes back to the period when Black was investigating the nature of heat.

52.

LAVOISIER, ANTOINE LAURENT (1743-1794)

LETTER French. Unpublished. The file contains two more letters dated 22.1.1781 (published in *Lychnos,* 1992, pp. 167-170) and 20.3.1791 (unpublished).
PLACE & DATE Paris, 14.11.1793.
DIMENSIONS 19.9 x 15.5 cms.

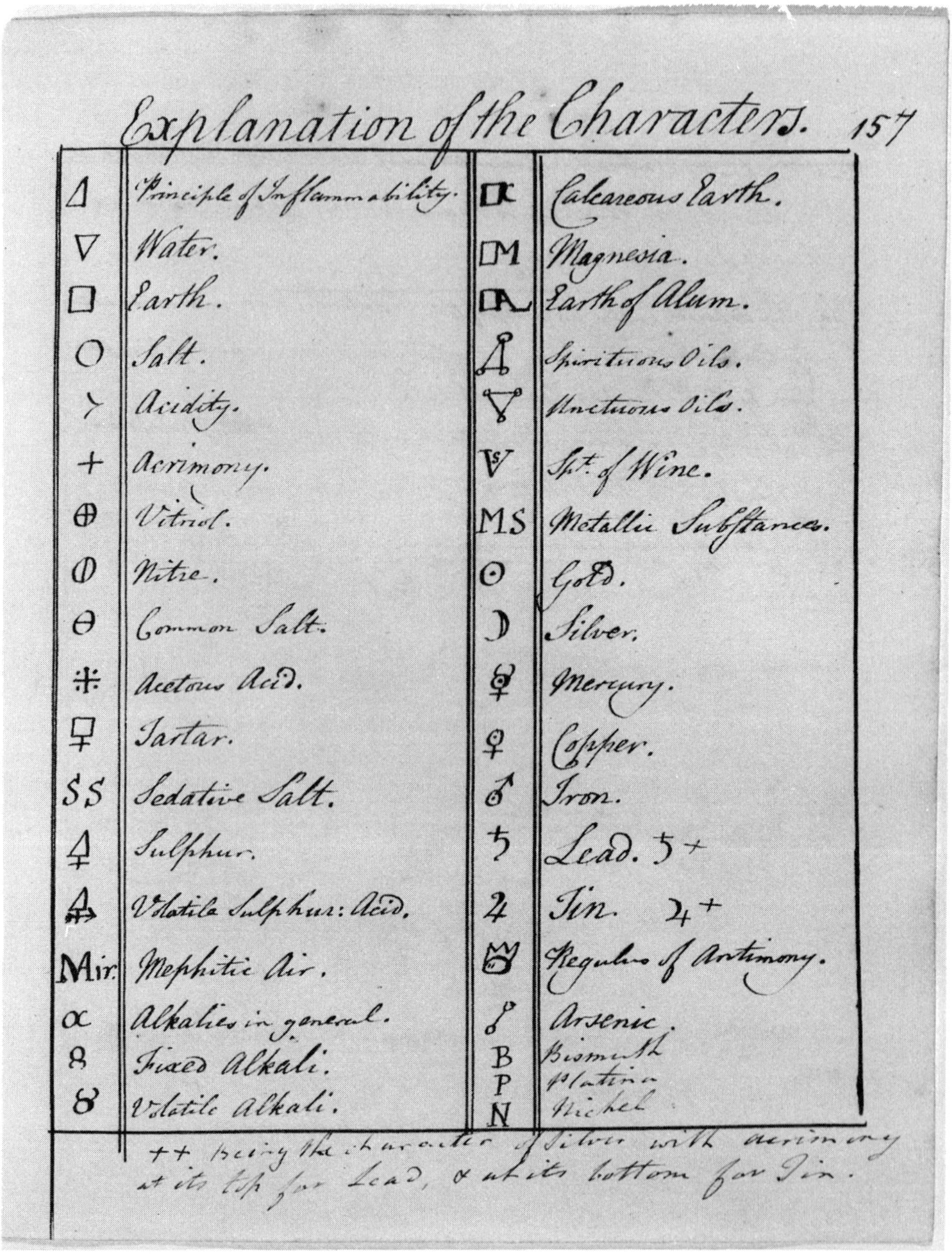

Explanation of the Characters. 157

△	Principle of Inflammability.	□C	Calcareous Earth.
▽	Water.	□M	Magnesia.
□	Earth.	□A	Earth of Alum.
○	Salt.	🜕	Spirituous Oils.
>	Acidity.	🜖	Unctuous Oils.
+	Acrimony.	▽S	Spt. of Wine.
⊕	Vitriol.	MS	Metallic Substances.
⦶	Nitre.	☉	Gold.
θ	Common Salt.	☽	Silver.
⁜	Acetous Acid.	☿	Mercury.
🜿	Tartar.	♀	Copper.
SS	Sedative Salt.	♂	Iron.
🜍	Sulphur.	♄	Lead. ♄+
🜍	Volatile Sulphur: Acid.	♃	Tin. ♃+
Mir.	Mephitic Air.	♁	Regulus of Antimony.
∝	Alkalies in general.	🜺	Arsenic.
8	Fixed Alkali.	B	Bismuth
8	Volatile Alkali.	P	Platina
		N	Nichel

++ being the character of Silver with acrimony at its top for Lead, & at its bottom for Tin.

Fig. 51: *Table of chemical symbols.*

FOLIATION 2 leaves.
SHELFMARK Waller 649 B:l.
HISTORICAL NOTE

Besides being the most authoritative chemist of the XVIIIth century, Lavoisier was a prominent public figure in the administration at the time of both the *Ancien Régime* and the French Revolution. This letter concerns Lavoisier's administrative responsibilities for French National Treasury; in it Lavoisier asks his relative, Parisis, to help him with some funds: "J'ay bien besoin que vous me procuries des fonds pour subvenir aux charges publiques qui augmentent de jour en jour. Il est dans mes principes de faire les plus grand efforts pour y subvenir et je ne le puis que sur mes revenus". Interestingly, Lavoisier was arrested a few days later for his former position as a Farmer General of the King and, after having been accused of gaining personal advantages from his tax collecting, he was judged guilty by the revolutionary Tribunal and executed on May 8, 1794.

53.

LAGRANGE, JOSEPH-LOUIS (1763-1813)

LETTER French. Unpublished?
PLACE & DATE Paris, 29.12.1793.
ADDRESSEE Jean-Baptiste Delambre.
DIMENSIONS 21 x 16 cms.
FOLIATION 2 leaves.
SHELFMARK Waller 649 B:1.
TRANSCRIPTION (Partial)

Paris 9 Nivose l'an 2 de
la repub. (29 10bre 1793)

Je suis chargé, Citoyen, de vous prevenir que le Comité de Salut public par une arreté du 3 nivos. a supprimé plusieurs membres de la Commission des poids et mesures, dans le nombre des quels vous etes compris. il est nécessaire en consequence que vous cessiez vos operations [...]
Je suis faché qu'une operation aussi importante pour les sciences soit interrompue et je désire plus que je n'espère, qu'elle puisse être continuée avec le même succès [...]

Votre Citoyen Lagrange

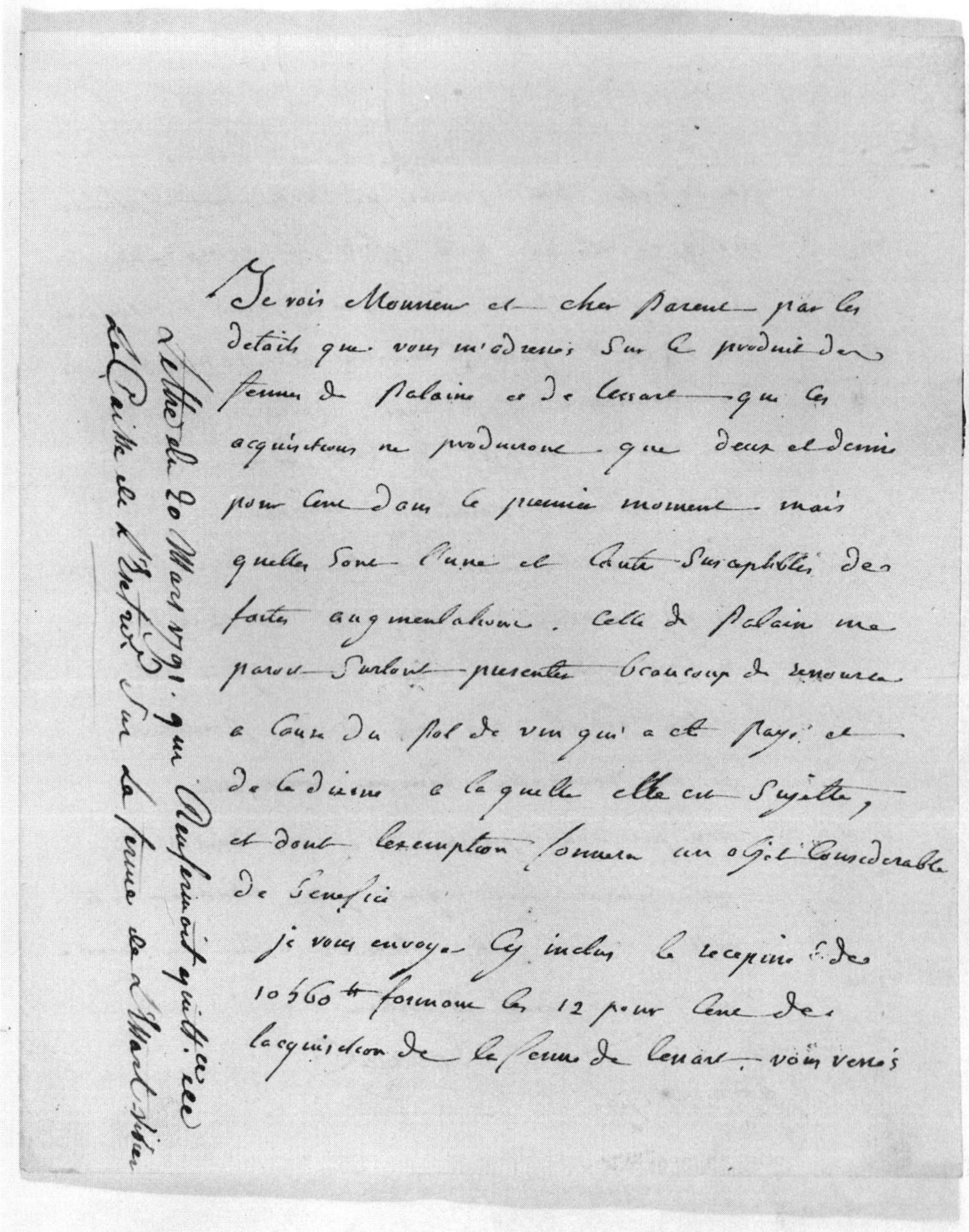

Fig. 52.

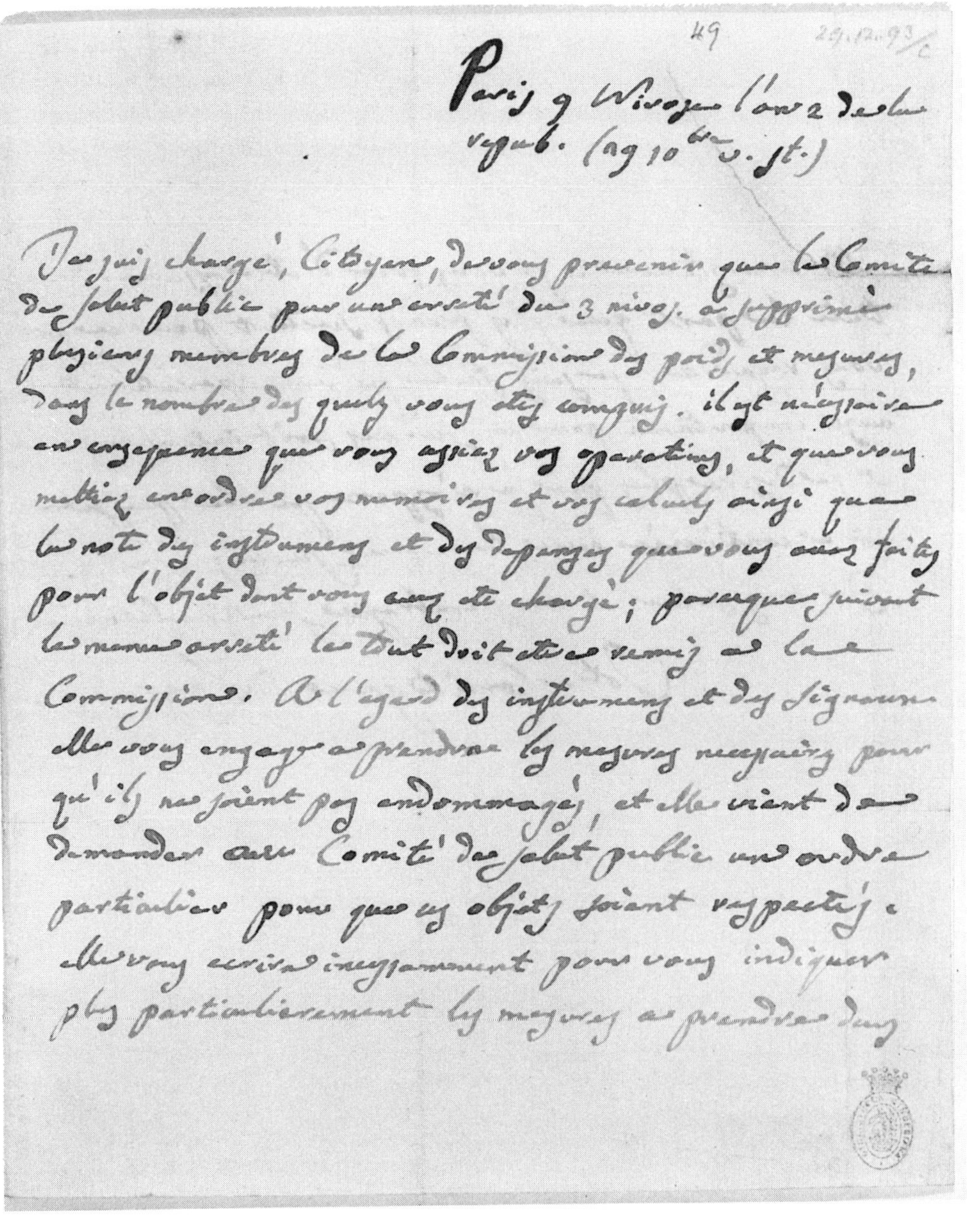
49 29.12.93/c

Paris 9 Nivose l'an 2 de la republ. (29 10bre v. st.)

Je suis chargé, Citoyen, de vous prevenir que le Comité de salut public par un arrêté du 3 nivos. a supprimé plusieurs membres de la Commission des poids et mesures, dans le nombre des quels vous êtes compris. il est nécessaire en conséquence que vous cessiez vos operations, et que vous mettiez en ordre vos memoires et vos calculs ainsi que la note des instrumens et des depenses que vous avez faites pour l'objet dont vous avez été chargé; parceque suivant le même arrêté le tout doit être remis à la Commission. A l'egard des instrumens et des signaux elle vous engage à prendre les mesures necessaires pour qu'ils ne soient pas endommagés, et elle vient de demander au Comité de salut public un ordre particulier pour que ces objets soient respectés. elle vous ecrira incessamment pour vous indiquer plus particulierement les mesures à prendre dans

Fig. 53.

HISTORICAL NOTE

On December 29, 1793, Prieur de la Côte d'Or persuaded the revolutionary Comité de Salut Public that several members of the Commission established to determine a new system of weights and measures had to be expelled. The victims of the purge were Borda, Laplace, Lavoisier, Coulomb, Brisson and Delambre. With this decision the revolutionary Government sanctioned the end of the last scientific association still in existence after the closure of the *Académie des sciences* and several other scientific institutions. The main purpose of these sanctions was to eliminate any trace of the old regime and to ensure that scientists shared the republican principles. When Lagrange wrote this letter Delambre was in the Loiret for the measurement of the meridian. The decimal system was soon after adopted by the Convention National but it was not until the end of the century that the work undertaken by Delambre and the other members of the Commission was effectively begun.

54.

MESMER, FRANZ ANTON (1734-1815)

LETTER — French. The file also contains two more letters dated 28.7.1799 and 25.2.1806, two autographs and two French official documents related to Mesmer's visit to France.

PLACE & DATE — Scaffhousen. 7.12.1797.

ADDRESSEE — M.Segretier.

DIMENSIONS — 22.9 x 18.1 cms.

FOLIATION — 2 leaves.

SHELFMARK — Waller 652 D:3.

TRANSCRIPTION — (Partial)

mon intention est, à mon arrivé le primtems prochain en France, de publier l'introduction à mon systeme que j'ai fini avec un avis au public qui sera accompaigné du second memoire que vous connoissez sur les effets du M[agnetisme] particulierement du somnambulisme. Je ne crois pas qu'on pourra être indifferent sur l'un et l'autre de ces deux objets.

HISTORICAL NOTE

After his failure in spreading his theory on magnetism in France in the 1780's, Mesmer persisted in his attempt to reach out to his French followers. In 1784 a commission was established to examine the scientific validity of Mesmer's claim on the healing power of magnetism. The members of the commission, Lavoisier, Bailly, Franklin, LeRoi and Bory concluded that Mesmer had abused the credulity and ignorance of his patients and that there was no experimental evidence of the existence of the magnetic fluid. After this negative assessment Mesmer was forced to leave Paris but he continued to keep in contact with his followers from his exile in Switzerland. The work mentioned in this letter is probably *Mémoire de F. A. Mesmer sur ses découvertes* (Paris, 1799). However, his hope that its reception would not leave his readers indifferent was not fulfilled.

55.

LAVATER, JOHANN GASPARD (1741-1801)

DRAWING Watercolour representing a wolf. In the file there are also four autograph letters and four autographs.

DATE 1793.

DIMENSIONS 14.1 x 11.4 cms.

SHELFMARK Waller 650 C:2.

HISTORICAL NOTE

Poet and theologian by education, Lavater founded the controversial science of physiognomy on a modern basis. Between 1775 and 1778 Lavater published his *Physiognomische Fragmente zur Beförderung der Menschenkenntniss und Menschliebe.* In this work Lavater explained the characters of man by their different physiognomies. This work, which had a resounding success and was translated into most European languages, raised a fierce controversy concerning its scientific validity.

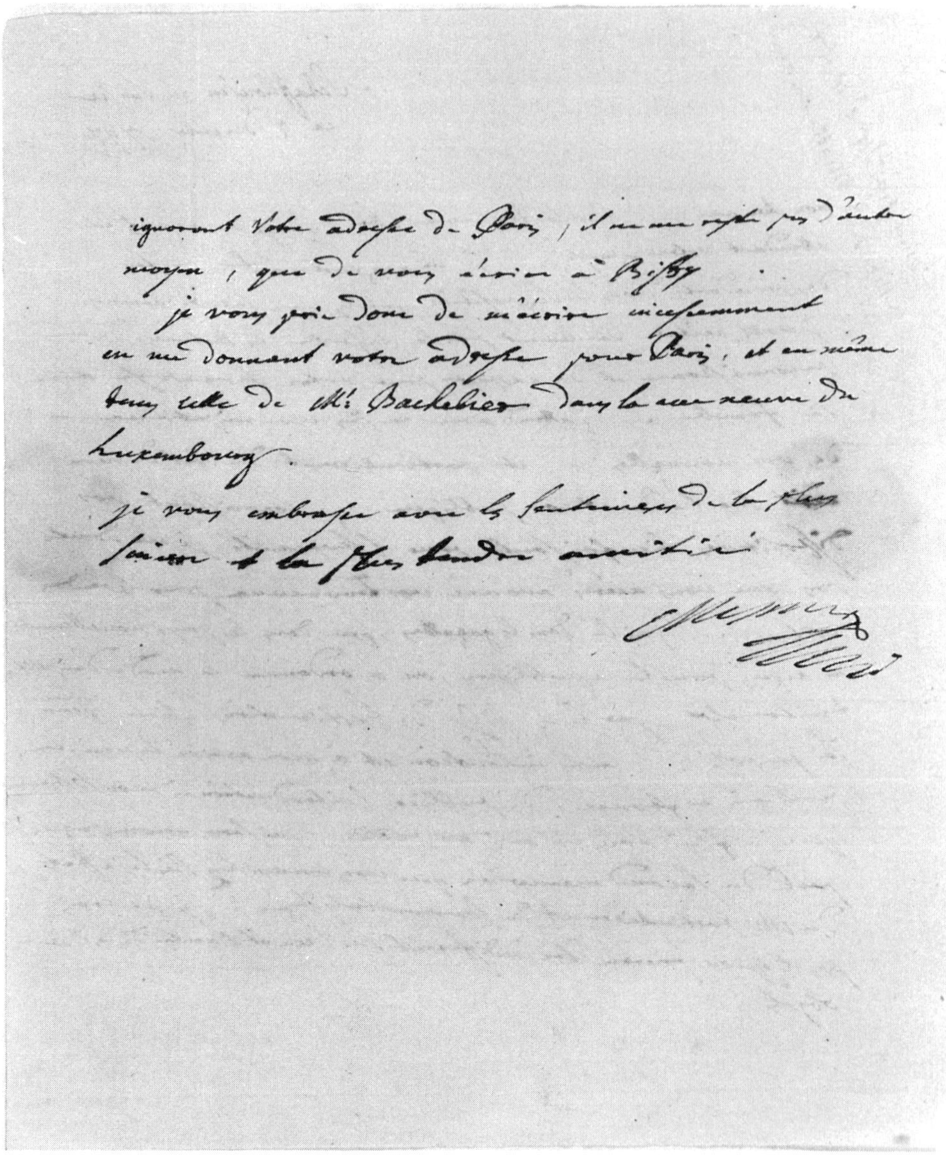

ignorant votre adresse de Paris, il ne me reste pas d'autre
moyen, que de vous écrire à Bissy
je vous prie donc de m'écrire incessamment
en me donnant votre adresse pour Paris, et en même
temps celle de Mr. Bachelier dans la rue neuve du
Luxembourg.
je vous embrasse avec les sentiments de la plus
sincère et la plus tendre amitié

Fig. 54.

Fig. 55.

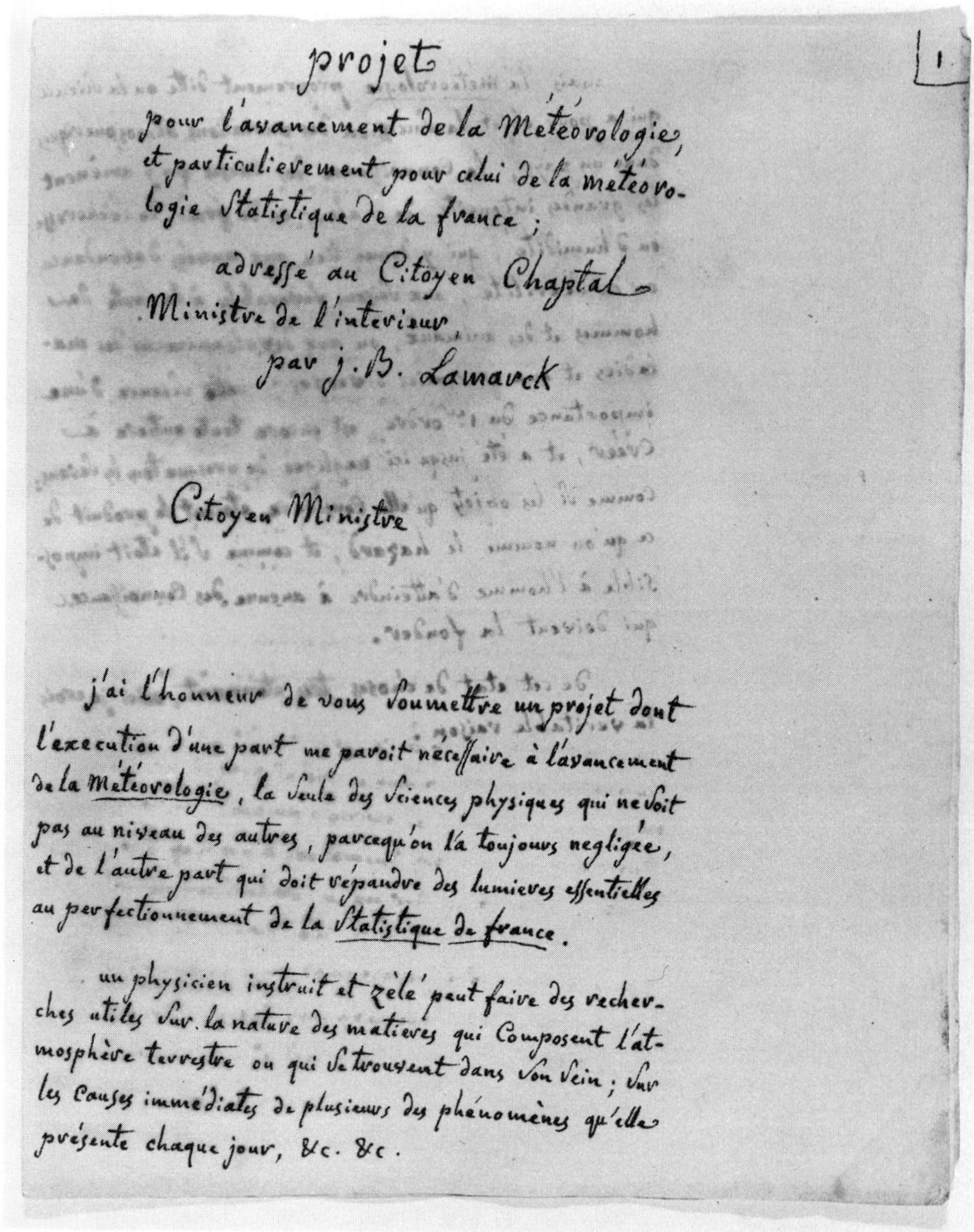

1

projet

pour l'avancement de la Météorologie, et particulièrement pour celui de la météorologie Statistique de la france;

adressé au Citoyen Chaptal
Ministre de l'intérieur

par j. B. Lamarck

Citoyen Ministre

j'ai l'honneur de vous soumettre un projet dont l'execution d'une part me paroit necessaire à l'avancement de la Météorologie, la seule des sciences physiques qui ne soit pas au niveau des autres, parcequ'on l'a toujours negligée, et de l'autre part qui doit répandre des lumieres essentielles au perfectionnement de la Statistique de france.

un physicien instruit et zélé peut faire des recherches utiles sur la nature des matieres qui composent l'atmosphère terrestre ou qui se trouvent dans son sein; sur les causes immédiates de plusieurs des phénomènes qu'elle présente chaque jour, &c. &c.

Fig. 56.

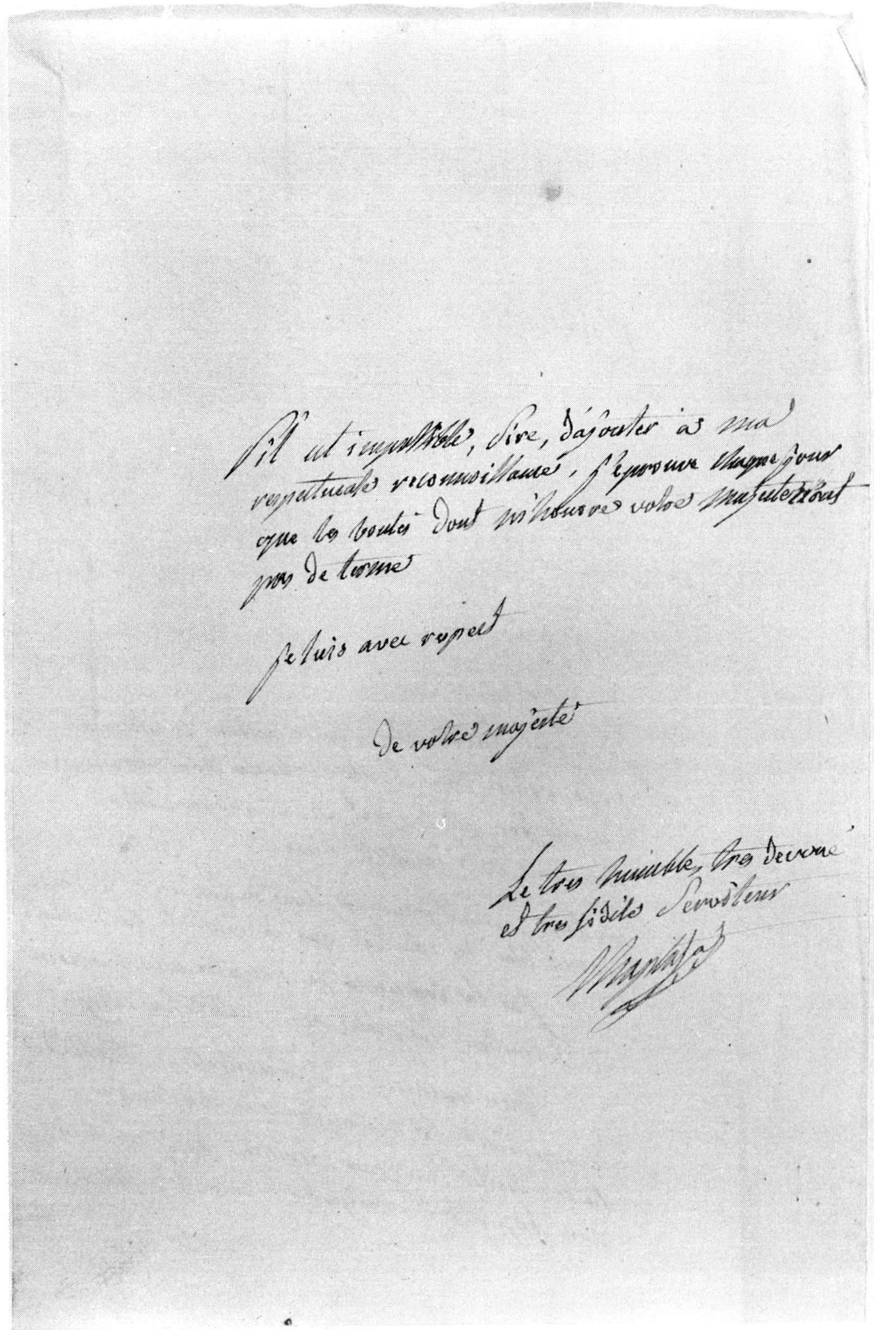

S'il est impossible, Sire, d'ajouter à ma respectueuse reconnaissance, j'éprouve chaque jour que la bonté dont m'honore votre Majesté n'a pas de terme.

Je suis avec respect

De votre majesté

Le très humble, très dévoué et très fidèle serviteur

Fig. 57.

56.

LAMARCK, JEAN BAPTISTE (1744-1829)

MANUSCRIPT French. The file also contains ten letters and a two-leaf Ms. bearing the title "Nottes relatives aux observations que le zoologiste peut faire à la mer et sur les côtes".

TITLE Projet pour l'avancement de la Météorologie et particulierement pour celui de la météorologie statistique de la France.

PLACE & DATE Paris, 20.8.1803.

DIMENSIONS 22.6 x 17.5 cms.

FOLIATION 11 leaves.

SHELFMARK Waller 649 B:1.

HISTORICAL NOTE

Dedicated to the Minister of the Interior, Jean-Antoine Chaptal, this Ms. is a proposal by Lamarck for an ambitious meteorological project. In it Lamarck is asking Chaptal to consider the possibility of creating a network of meteorological observations throughout the French territories and to coordinate the data by following his new statistical method.

"Je pense", Lamarck continues, "qu'il seroit tres avantageux pour l'avancement de la Météorologie, pour amener cette Science jusqu'aujourdhui généralement negligée, au niveau des progrès des autres sciences physiques, et particulierement pour fonder la Météorologie Statistique de la France; de publier Régulierement tous les mois les observations météorologiques recueillies simultanément dans un certain nombre de points du territoire français ou l'on observe".

The principal goal of this project was to collect enough data to allow accurate weather forecasts and measurements of atmospheric changes. In Lamarck's view, however, the main concern was to see his statistical method generally adopted. Chaptal, who probably thought that such a project could have a positive effect on the Napoleonic Wars, initially supported the project. However, the difficulties in coordinating different methods of collecting data with Lamarck's approach created insurmountable problems to its realization and after a few years the project was finally abandoned.

57.

CHAPTAL, JEAN ANTOINE CLAUDE (1756-1832)

LETTER: French. Unpublished. The file contains another letter addressed to Dumeril dated 13.3.1803 and three letters written by Chaptal as Minister of the Interior of France.

PLACE & DATE: Paris, 1804 (le11e de l'an 13).

ADDRESSEE: Napoleon I.

DIMENSIONS: 31.5 x 20.5 cms.

FOLIATION: 1 leaf.

SHELFMARK: Waller 648 E:2.

HISTORICAL NOTE

Letter to Napoleon in which Chaptal accepts his nomination as *Trésorier du Sénat* (Treasurer of the Senate) announced on October 17, 1804. Chaptal also promises Napoleon to perform his duties with distinction. Chaptal was a distinguished chemist and one of the first to accept Lavoisier's theory of oxygen in Europe. He was also among the first to apply the late discoveries made in chemistry to manufactures and agriculture. When Napoleon came to power, Chaptal left his scientific career to become a prominent politician. Member of the *Conseil d'État* in 1799 and of the Senate, Chaptal was elected by Napoleon as Minister of the Interior in 1801, a position which he retained until 1804. During this period Chaptal supported and financed several scientific projects (see item No. 56).

58.

CHAPPE, CLAUDE (1763-1805)

TELEGRAPHIC DISPATCH: French. The file also contains an official report on the telegraph signed by Chappe and dated 1802.

SENDER: General of the French army. Signed by Chappe.

PLACE & DATE: d'Huminique, 8.4.1800.

ADDRESSEE: French Minister of War.

DIMENSIONS: 23.2 x 17.6 cms.

FOLIATION: 1 leaf.

ILLUSTRATED: A vignette with Mercury flying through the clouds bearing a telegraph.

SHELFMARK: Waller 648 E:2.

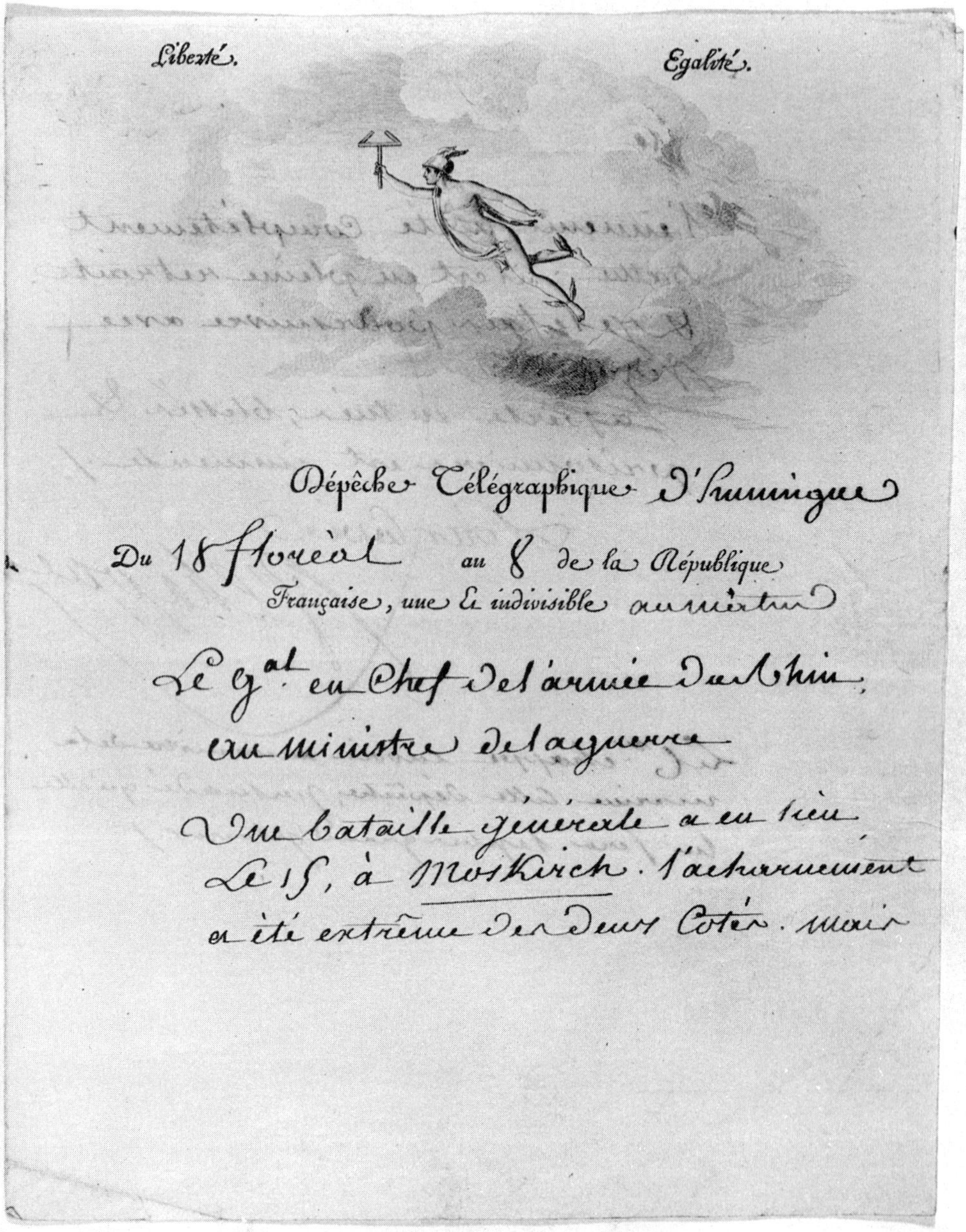

Liberté. Egalité.

Dépêche Télégraphique d'huningue

Du 18 floréal au 8 de la République Française, une et indivisible

Le Gal. en Chef de l'armée du Rhin, au ministre de la guerre

Une bataille générale a eu lieu le 15, à Moskirch. l'acharnement a été extrême des deux Côtés. mais

Fig. 58.

TRANSCRIPTION

général en chef de l'armée du Rhin au Ministre de la guerre

d'Huminique 18 Floréal au VIII au matin

Une Bataille générale a eu lieu le 15 à Maskirch. L'acharnement a été extreme de deux côtes. mais l'ennemi a été complètement battu; il est en pleine retraite et je le fais poursuivre avec vigueur. Sa perte en tués, blessés et prisonniers est immense.

Sous copie
Chappe

HISTORICAL NOTE

At the beginning of the 1790's the French physicist, Claude Chappe, invented an optical telegraph system capable of transmitting information over a distance of 15 kilometers. Aware of its possible application in warfare, the French National Convention encouraged Chappe to make further research in this field and by 1798 two telegraphic lines had been completed on French territory. During the Napoleonic Wars the telegraph provided the generals with prompt communication on the manoeuvres of the enemy and ensured the Minister in Paris with rapid up-to-date information on changes at the front lines.

59.

KRAUSE, CARL FRIEDRICH THEODOR (1798-1868)

MANUSCRIPT Latin. Unpublished.

TITLE Synopsis icone illustrata nervorum systematis gangliosi in capite homini.

PLACE & DATE Hanover, 1839.

DIMENSIONS 43 x 28 cms.

FOLIATION 4 leaves.

ILLUSTRATED

SHELFMARK Waller 653 F:2.

HISTORICAL NOTE

The Ms. is a description of the nervous system of the human head. A pupil of Blumenbach, Krause practised as a physician in the criminal department of Hanover for more than half a century (until 1868). He published several works on anatomy, pathology, and physiology. The origin of this manuscript is described by his son W. Krause as follows: "Es war zugelich ein

Fig. 59.

Zeichnen der Dankbarkeit, dass zu Stieglitz's Doctor-Jubiläum eine lateinishce Gratulationsschrift erschein [...] In dieser mit einer sehr bekannten und oft reproducirten Tafel ausgestatten Monographie zeigte sich die ganze Meisterschaft der feinsten anatomischen Technik" (from W. Krause, *Die Motorischen Endplatten der quer-gestreiften muskelfasern,* Hanover, 1869).

60.

BERNARD, CLAUDE (1813-1878)

LETTER French. Unpublished. The file contains six autograph letters.
PLACE & DATE Paris, 22.3.1854.
ADDRESSEE Napoleon III.
DIMENSIONS 32 x 20.2 cms.
FOLIATION 1 leaf.
SHELFMARK Waller 649 A:1.
TRANSCRIPTION

A sa Majesté
L'Empereur Napoleon III.
Sire,

Je prie votre Majesté de me permettre de lui exprimer ma profonde reconnoissance pour l'honneur qu'elle m'a fait en m'appelant à la chaire de Physiologie Générale crée à la Faculté des Sciences de Paris.

Cette nomination m'impose de grands devoirs envers la science et l'enseignement. Je ferai tous mes efforts pour les remplir et pour me montrer digne de la haute bienveillance dont l'Empereur m'a honoré.

Je suis avec le plus profond respect,
Sire,
de votre Mejesté le très-humble et très obéissant serviteur.
Cl. Bernard.
Professeur à la faculté des sciences de Paris

22 Mars 1854.

HISTORICAL NOTE

After receiving a doctoral degree in zoology in 1853, Claude Bernard was appointed to the chair of General Physiology, created for him by the French Government in 1854. During the period 1854-1860 Bernard published several pioneering works devoted to experimental physiology which soon became the most authoritative studies in this field throughout Europe.

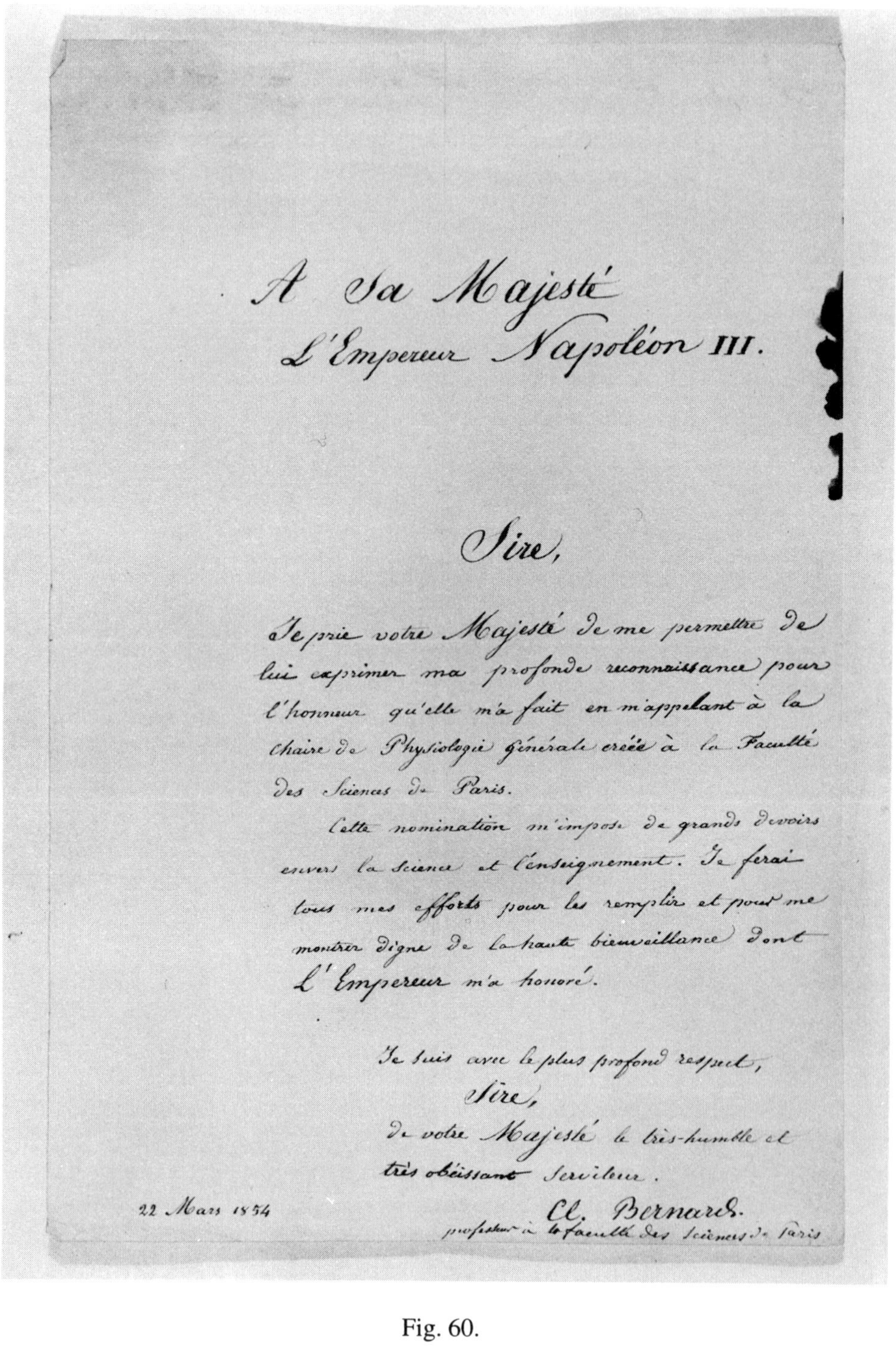
A Sa Majesté
L'Empereur Napoléon III.

Sire,

Je prie votre Majesté de me permettre de lui exprimer ma profonde reconnaissance pour l'honneur qu'elle m'a fait en m'appelant à la chaire de Physiologie générale créée à la Faculté des Sciences de Paris.

Cette nomination m'impose de grands devoirs envers la Science et l'enseignement. Je ferai tous mes efforts pour les remplir et pour me montrer digne de la haute bienveillance dont l'Empereur m'a honoré.

Je suis avec le plus profond respect,
Sire,
de votre Majesté le très-humble et très obéissant Serviteur.

22 Mars 1854

Cl. Bernard.
professeur à la Faculté des Sciences de Paris

Fig. 60.

61.

DARWIN, CHARLES (1809-1882)

LETTER — English. Unpublished. The file also contains a photograph of Darwin bearing his autograph and four autograph letters.

PLACE & DATE — Down Beckenham, Kent, 10.12.1879.

ADDRESSEE — --

DIMENSIONS — 20.1 x 12.5 cms.

FOLIATION — 2 leaves.

SHELFMARK — Waller 648 A:2.

TRANSCRIPTION

Down Beckenham Kent

Dec. 10 1879

Dear Sir,

I have no objection to express my opinion on the subject of vivisection. I detest cruelty, & any physiologist who in investigation causes suffering which can possibly be avoided deserves universal execration.

But I am convinced that physiology, which cannot progress without experiments on living animals, is one of the most important of the sciences, & that it will hereafter confer inestimable benefits on mankind.

Whether or not it has already improved the practice of medicine, is a question of comparative insignificance; for the history of science shews that benefits to man are generally the indirect result of the search for truth. It is my conviction that he who stops or delays the progress of physiology will hereafter be looked at as the greatest enemy to the sacred cause of true humanity.

If physiologists are reckless in their experiments, which as far as I can discover has not been the case in England, assuredly they ought to be checked & punished by public opinion or the law.

I remain
dear Sir
yours faithfully
Charles Darwin

P.S. It is not probable that you will circulate this letter, but if you do, I appeal to your honour to print the whole of it.

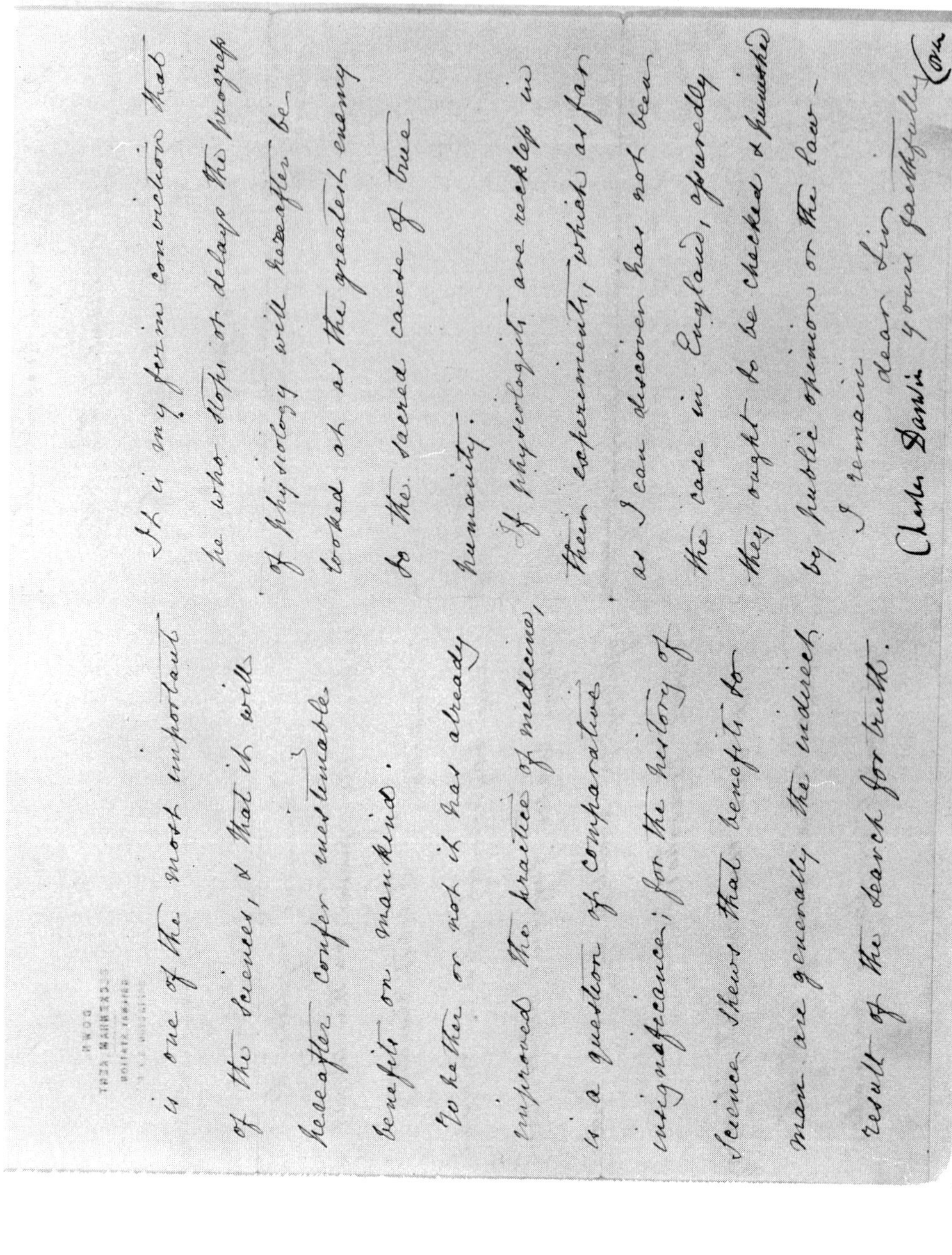

is one of the most important
of the Sciences, & that it will
hereafter confer inestimable
benefits on mankind.
Whether or not it has already
improved the practice of medicine,
is a question of comparative
insignificance; for the history of
Science shews that benefits to
man are generally the indirect
result of the search for truth.

It is my firm conviction that
he who stops or delays the progress
of physiology will hereafter be
looked at as the greatest enemy
to the sacred cause of true
humanity.

If physiologists are reckless in
their experiments, which as far
as I can discover has not been
the case in England, assuredly
they ought to be checked & punished
by public opinion or the law.

I remain
dear Sir
yours faithfully
Charles Darwin

Fig. 61 a.

Fig. 61 b.

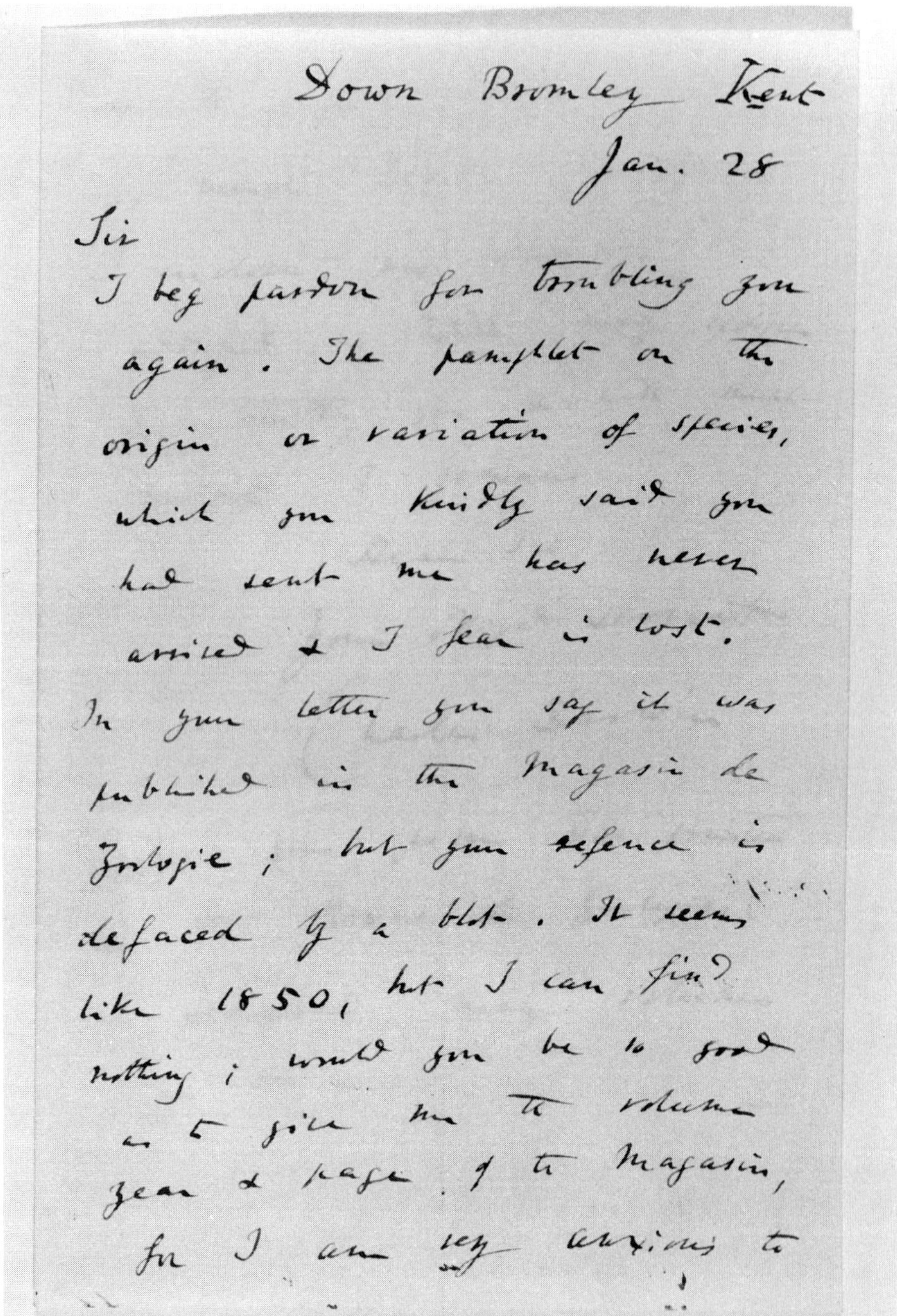
Down Bromley Kent
Jan. 28

Sir
I beg pardon for troubling you again. The pamphlet on the origin or variation of species, which you kindly said you had sent me has never arrived & I fear is lost. In your letter you say it was published in the Magasin de Zoologie; but your reference is defaced by a blot. It seems like 1850, but I can find nothing; would you be so good as to give me the volume, year & page of the Magasin, for I am very anxious to

Fig. 62.

62.

DARWIN, CHARLES (1809-1882)

LETTER + ADDRESSED ENVELOPE. Another letter by Darwin is kept in one of the 14 volumes of autographs.

PLACE & DATE Down Bromley, Kent, 28.1.1860.

ADDRESSEE Isodore Geoffroy Saint Hilaire.

DIMENSIONS 9.9 x 12.3 cms.

FOLIATION 2 leaves.

SHELFMARK Waller Delikatesskaps 1.

TRANSCRIPTION

Down Bromely Kent
Jan. 28

Sir
I beg pardon for troubling you again. The pamphlet on the origin or variation of species, which you kindly said you had sent me has never arrived & I fear is lost.
In your letter you say it was published in the Magasin de Zoologie; but your reference is defaced by a blot. It seems like 1850, but I can find nothing; would you be so good as to give me the volume year & page of the Magasin, for I am very anxious to see the paper. To save you as much trouble as possible I enclose an envelope addressed. With many apologies for troubling you & with much respect, I remain

Dear Sir
your obliged servant
Charles Darwin

Perhaps your paper was printed in the Revue de Zoologie. I searched many volumes in vain.

HISTORICAL NOTE

In 1859 I. Geoffroy Saint Hilaire published a paper *(Résumé des vues sur l'espèce organique)* in which he vindicated the originality of French investigations in the field of the variation of the species, namely those made by Buffon, Lamarck, and E. Geoffroy Saint Hilaire. The paper was intended to show that Darwin's *Origin of Species* (pub.1859) was not as original as was claimed by the author. It is therefore natural to find Darwin extremely anxious to read the paper, which in fact did not have the effect hoped for by Geoffroy. There are 25 letters by I. Geoffroy Saint Hilaire (many of which are addressed to his teacher Constant Duméril) in Waller 649 A:1.

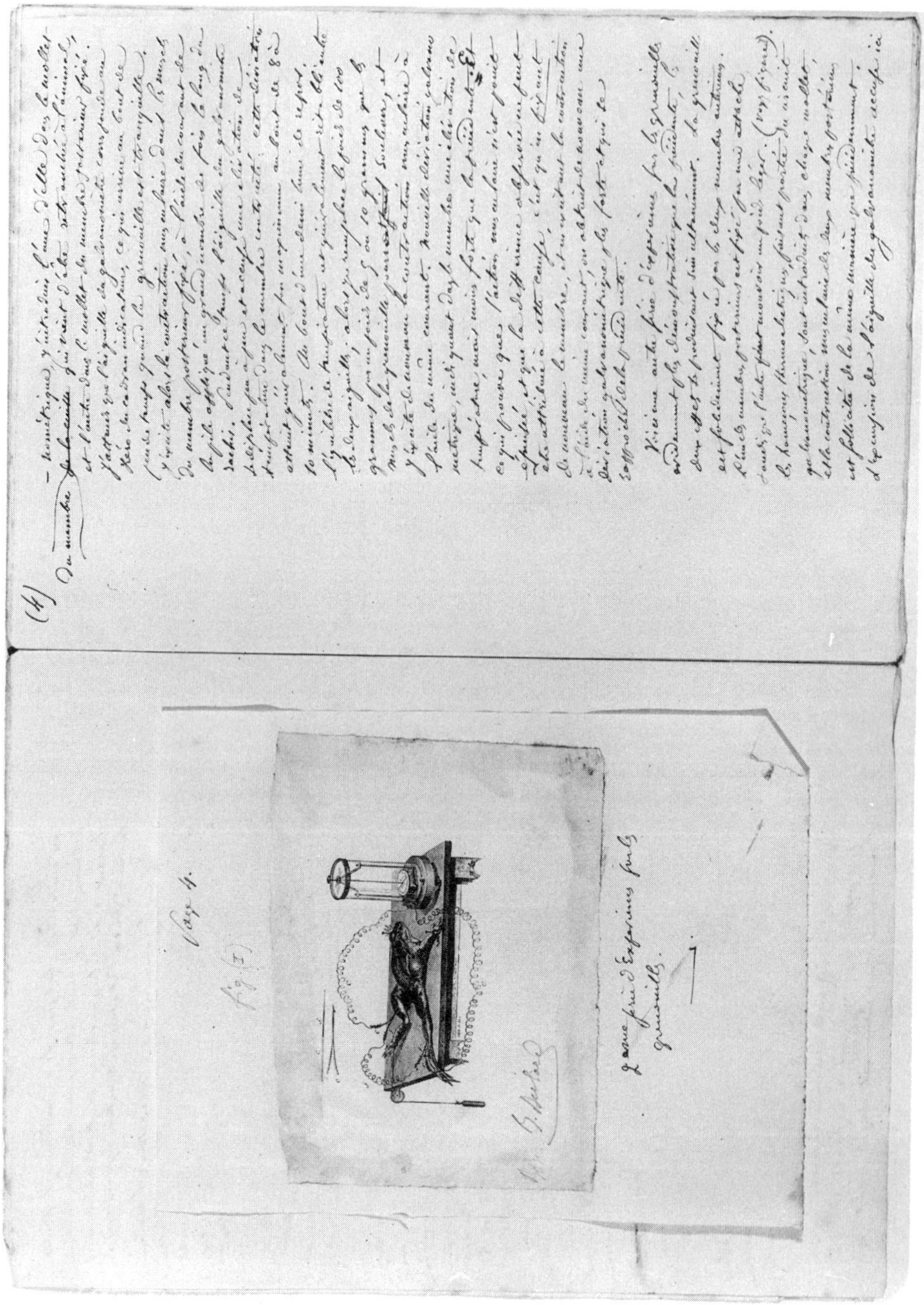

Fig. 63.

63.

BÉCLARD, JULES AUGUSTIN (1817-1887)

MANUSCRIPT	French. In the same file there are eight autograph letters by Béclard.
TITLE	De la chaleur produite pendant le travail de la contraction musculaire.
DATE	approx. 1850.
DIMENSIONS	26 x 19.3 cms.
FOLIATION	17 leaves + 1 copperplate and 1 drawing.
ILLUSTRATED	
UNBOUND	
SHELFMARK	Waller 649 A:l.
HISTORICAL NOTE	

Memoir aimed at demonstrating the production of heat during the contraction of muscles. The memoir examines critically the opinions of those scientists such as Becquerel, Helmholtz, and Dreschet who raised doubts about this hypothesis. This memoir was presumably read before the *Académie Impériale de Médicine de Paris* of which Béclard was a member.

64.

MOLESCHOTT, JACOB (1822-1893)

MANUSCRIPT	German. Published.
TITLE	Der Kreislauf des Lebens, Physiologische Antworten auf Liebig's Chemische Briefe. von Jac. Moleschott.
PLACE & DATE	Heidelberg, 3.4.1852.
DIMENSIONS	23 x 18 cms.
FOLIATION	8 + 539 leaves.
SHELFMARK	Waller 653 D:2.
HISTORICAL NOTE	

Original manuscript of one of the most famous and controversial works on physiology published during the XIXth century. It is possible in fact that Moleschott's work was published in 1852 from this Ms. If compared with the first edition of *Der Kreislauf des Lebens* the text of the manuscript is identical to the printed version with the exception of the first four pages of the Ms. introduction which were not published.

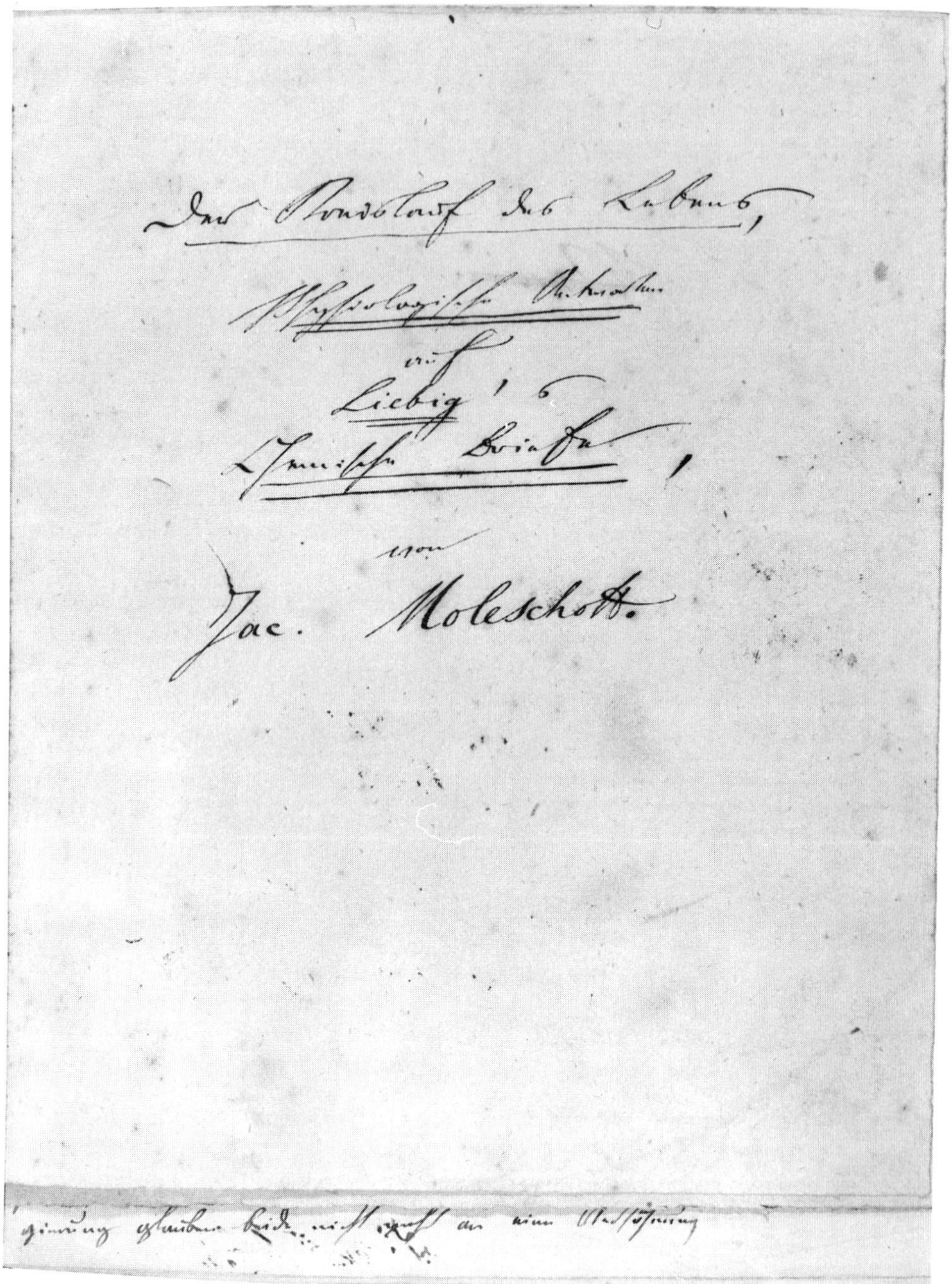
Der Kreislauf des Lebens,
Physiologische Antworten
auf
Liebig's
Chemische Briefe,
von
Jac. Moleschott.

Fig. 64.

In this work Moleschott discusses critically the spiritualistic philosophy of science outlined by the chemist Justus von Liebig in his popular work *Chemische Briefe*. Being a fervent materialist, Moleschott claimed that biological phenomena could be reduced to the constant "exchange of matter" (*Stoffwechsel*) occurring in nature. The scientific background of this belief was rather simplistic but Moleschott's passionate tone ensured his work a tremendous success. Moleschott's materialistic vision of biology is evidenced on fol. 35 of the Ms. where he wrote:
"Was der Mensch ausscheidet, ernärt die Pflanze. Die Pflanze verwandelt die Luft in feste Bestandtheile und ernärt das Thier. Raubthiere von Pflanzenfressen und selbst eine Beute des Todes zu werden und neues reimendes Leben in der Pflanzenwelt zu verbreiten.
Diesen Austauch des Stoffes hat man den Namen Stoffwechsel gegeben. Man spricht das Wort mit Recht nicht ohne ein Gefühl der Verehrung. Den wie der handel die Seele ist der Verkehrs, so ist das ewige Kreisen des Stoffs die Seele der Welt".

65.

VIRCHOW, RUDOLF (1821-1902)

MANUSCRIPT German. Published. The file contains another 20-leaf manuscript entitled "Über Cavernose (erectile) Geschwülste und Teleangiectasien". Two letters by Virchow are in Waller 652 C.

TITLE Cellularpathologie.

DATE 1855.

DIMENSIONS 21.5 x 16.9 cms.

FOLIATION 27 leaves.

SHELFMARK Waller 652 D:3.

HISTORICAL NOTE

This is the original manuscript version of one of the most important papers written in cellular pathology. In this pioneering work Virchow believed that the microscopic investigation of cells could provide medicine with the explanatory key for cancer and other human diseases. Virchow's motto *omnis cellula e cellula* eptomises his innovative approach to physiology. The Ms. was published by Virchow in 1855 in the *Archiv für Patologische Anatomie und Physiologie und für Klinische Medicin* (vol. 8, p. 39), a journal which was founded by Virchow himself in 1845.

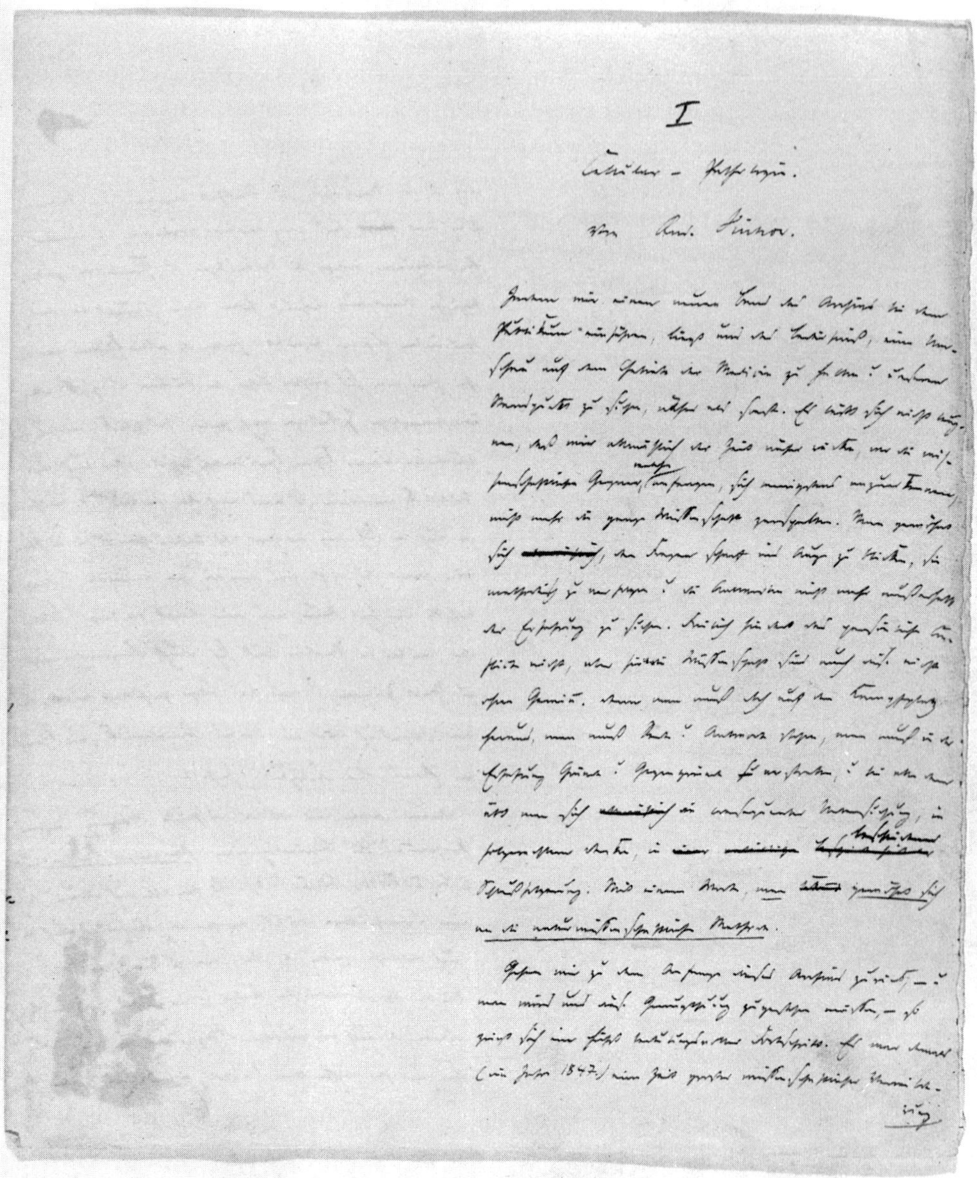

Fig. 65.

In this memoir Virchow rejected the "dogmatic" or transcendental concerns of the traditional pathological medicine and urged an entirely new approach centered on the systematic use of the microscope. For this purpose he felt the need for "eine wissenschaftliche Mikroscopie" and made an interesting analogy between biology and astronomy ("wie die Biologie ist mikroskopisch, so ist die Astronomie Teleskopisch" p. 16). As in astronomy it was impossible to make any useful observations without the help of the telescope, so in biology and pathology the use of the microscope was the only means of investigation. The emphasis on the importance of the microscope in biological research was underlined by Virchow's research on the pathological alterations of cells. It was in this paper that Virchow set forth his claim concerning the central role played by the cell in biology in a definitive form and wrote: "Ich formulire die Lehre von der patologischen Generation, von der Neoplasie in Sinne der Cellularpathologie einfach: *Omnis cellula e cellula*".

66.

PASTEUR, LOUIS (1822-1895)

LETTER French. Unpublished. There are three other letters by Pasteur in the same file

PLACE & DATE Arbois, 1.10.1878.

DIMENSIONS 20.5 x 13 cms.

FOLIATION 1 leaf.

SHELFMARK Waller 652 D:2.

TRANSCRIPTION (Partial)

Letter in which Pasteur intended to provide a refutation of a posthumous work by Claude Bernard in which he declared:
Je montrerai que ce grand physiologiste s'est trompé sur tous les points abordés. Il est fort intéressant de le suivre et de retrouver le chemin qui l'a conduit à l'erreur. On a cru m'embarasser par cette publication posthume qu'il n'eut jamais publié. En realité on aura rendu un très mauvais service a la mémoire de mon illustre ami et à moi on m'aura fourni l'occasion d'avoir raison contre un observateur trompe. Si je voyais dans la science et son progrès autre chose qu'elle même et l'honneur de mon pays je m'enorgueillirais de la circonstance qui m'a été offerte de défendre une fois de plus mes travaux...

HISTORICAL NOTE

In July 1878 Marcelin Berthelot published in the *Revue Scientifique* some posthumous notes by the French physiologist Claude Bernard in which he criticized Pasteur's theory of fermentation and claimed that he had found a method capable of producing alcoholic fermentation independently of living yeast. Soon after this letter was written Pasteur published his *Examen critique d'un Écrit posthume de Claude Bernard sur la fermentation,* (Paris, 1878) in which he demolished the experiments of Bernard.

67.

STRINDBERG, AUGUST (1849-1912)

LETTER French. Unpublished.
PLACE & DATE Paris, 27.2.1895.
ADDRESSEE [André Dubosc] general editor of *Science Française.*
DIMENSIONS 17.4 x 10.7 cms.
FOLIATION 1 leaf.
SHELFMARK Waller 650 F:2.
TRANSCRIPTION

Monsieur

Je suis désolé que l'apreté du style dans ma lettre vous a blessé; et je m'excuse en vous confiant que la chimie m'a coûté la bonheur de ma vie, presque mon existence.

——

Pour l'article que vous aviez demandé à la Science Française, je l'ai livré à M. Obolski sous condition qu'il serait imprimé cette termine. Avant tout je vous prie de ne rien couper, et je tiens le plus aux lignes renfermant les reflexions sur le 120° qui decident de la destinée du Soufre.

Je serai très heureux si vous voudriez m'accorder un rendez-vous, car j'ai à vous dire entre autre que je vous suis bien obligé, votre article ayant fait connoître mes idées.
Avec mes sentiments les plus distinguées

August Strindberg
12 rue de la grande Chaumière
Paris (Montparnasse)
27 Février 1895

HISTORICAL NOTE

Strindberg became interested in alchemy during the years 1895-1902. He contributed to the French alchemical journal *L'Hyperchimie* which was edited by his friend, J. Castelot, and he was co-founder and honorary member of the *Société Alchimique de France*. As is testified in this letter Strindberg's alchemical interests were an important part of his intellectual background. His involvement in alchemy was so deep that after the death of Alfred Nobel and the inauguration of the Nobel Prize he wrote to Castelot and declared: "Par là je me figure que les adhérents de votre Société alchimique seraient bien aptes à concourir pour le prix Nobel ...
Maintenant, le prix de chimie sera jugé par l'Académie des Sciences de Suède, et le chimiste là, le célèbre explorateur M. le baron Nordenskiold, le quel j'ai l'honneur d'appeler ami, connaît bien nos travaux de Chimie unitaire, et je pourrais vous dire qu'il n'est pas l'adversaire, sans en etre l'adhérent déclarée".
Strindberg's unitarian chemistry centered on the alchemical belief in the transmutation of matter. He claimed in fact that "les corps n'existent que par nombres et qualités. Les corps ne sont ni simples ni composées".

In spite of his constant effort to spread his ideas in scientific circles and academies, the response to Strindberg's chemistry was very negative and as early as 1898 Strindberg gave up his research. Useful references to the background of this letter and, more generally, to Strindberg's alchemical ideas can be found in *Bréviarie Alchimique. Lettres d'Auguste Strindberg à Jollivet Castelot,* (Paris, 1912) and in Strindberg's *Introduction à une chimie unitaire,* (Paris, 1895). *(Bibliotheca Walleriana,* nos. 11248-11249.)

68.

CURIE-SKLODOWSKA, MARIE (1867-1934)

HISTORICAL NOTE	French. Unpublished. The file contains three more autograph letters.
PLACE& DATE	Paris, 7.12.1908.
ADDRESSEE	G. Clemenceau (Prime Minister of France).
DIMENSIONS	21.2 x 13.5 cms.
FOLIATION	2 leaves.

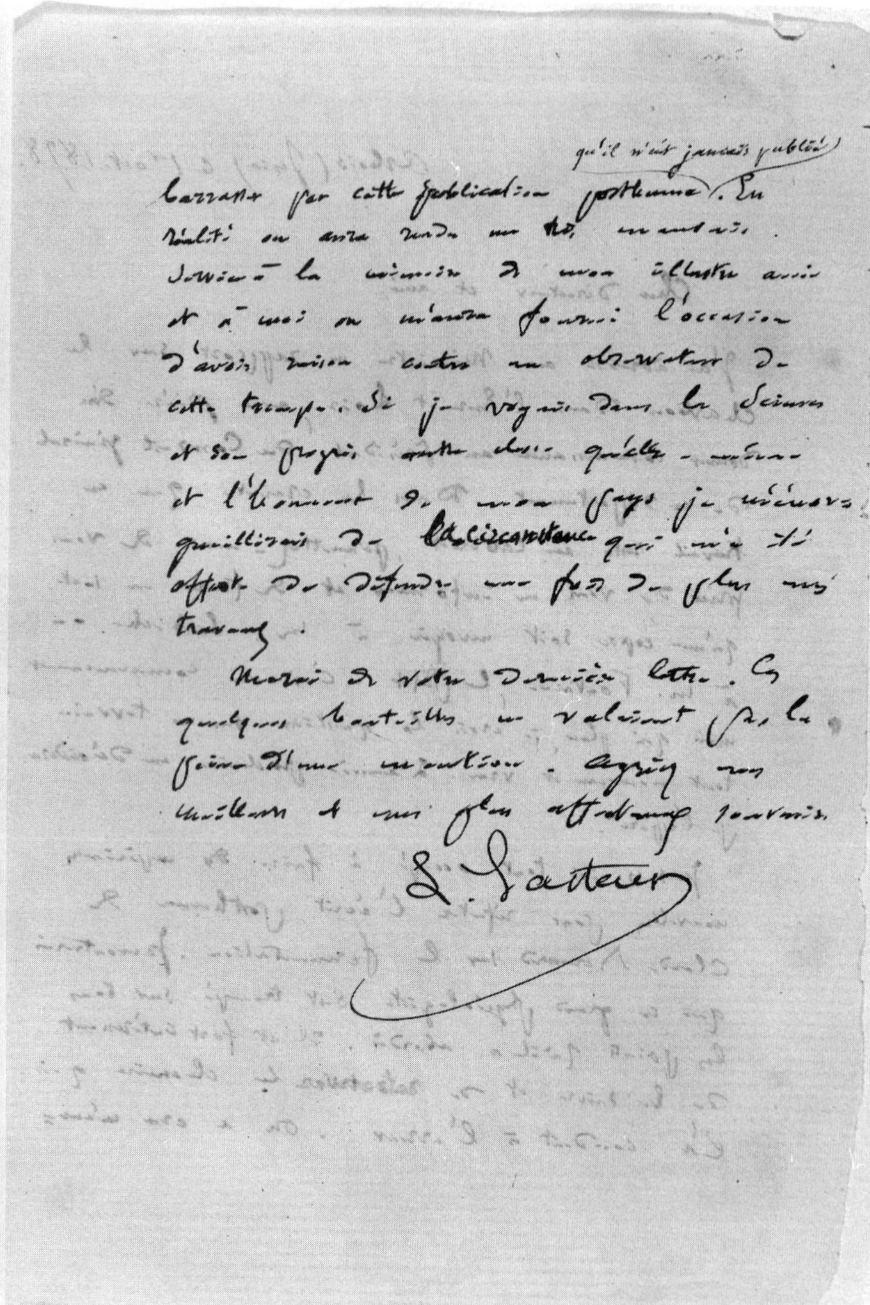

caractère par cette publication posthume qu'il n'eût jamais publiée. En réalité on aura rendu un très mauvais service à la mémoire de mon illustre ami et à moi on m'aura fourni l'occasion d'avoir raison contre un observateur de cette trempe. Si je regardais dans la Science et ses progrès autre chose qu'elle-même et l'honneur de mon pays je m'enorgueillirais de la circonstance qui m'a été offerte de défendre une fois de plus mes travaux.

Merci de votre dernière lettre. Ces quelques bouteilles ne valaient pas la peine d'une mention. Agréez mes meilleurs et mes plus affectueux souvenirs.

L. Pasteur

Fig. 66.

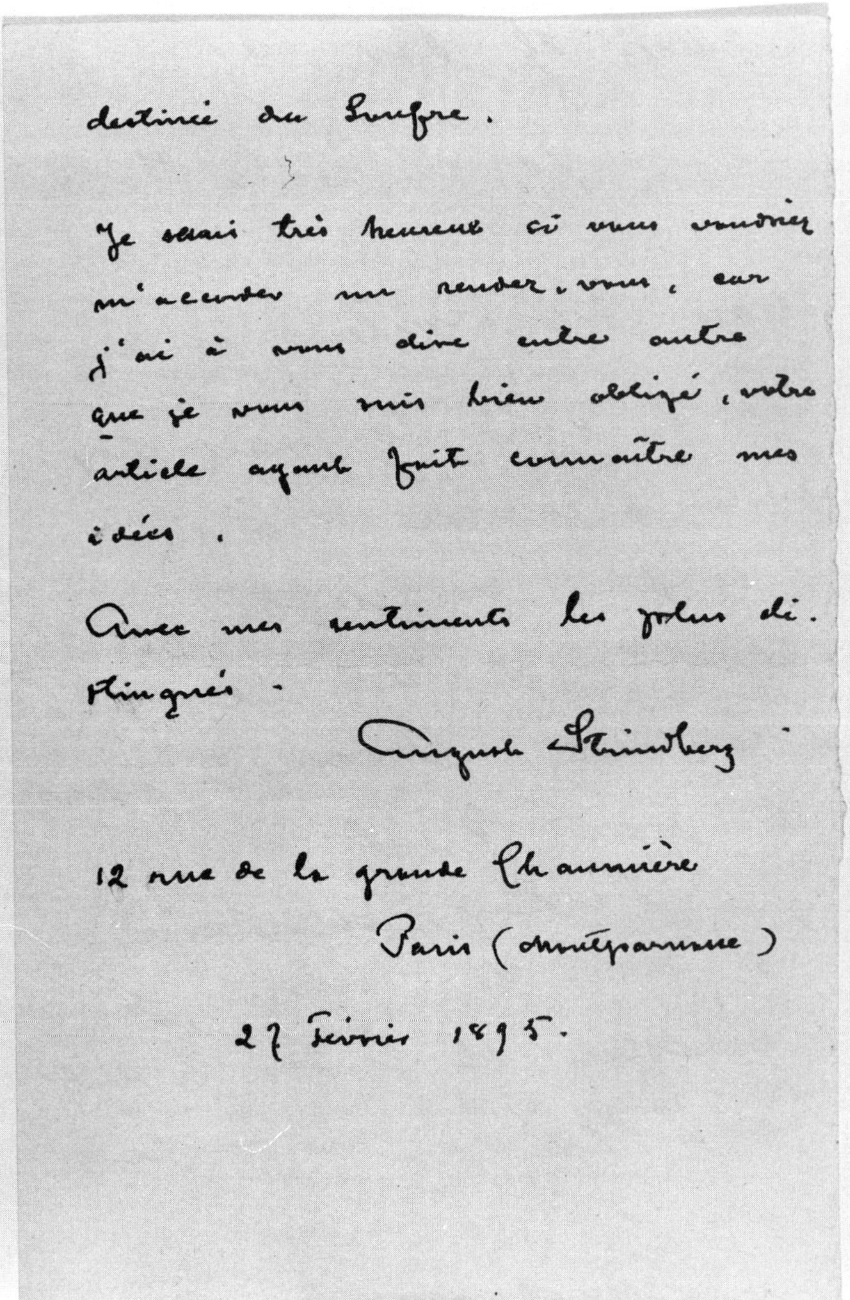

destinée du Soufre.

Je serai très heureux si vous voudriez m'accorder un rendez-vous, car j'ai à vous dire entre autre que je vous suis bien obligé, votre article ayant fait connaître mes idées.

Avec mes sentiments les plus di-stingués.

August Strindberg.

12 rue de la grande Chaumière
Paris (Montparnasse)

27 Février 1895.

Fig. 67.

SHELFMARK Waller 652 D:2.

TRANSCRIPTION

7 décembre 1908

Monsieur le Président du Conseil,

Puisque vous avez bien voulu vous interesser à l'affaire d'achat de minerai de radium pour mon laboratoire, je prends la liberté de vous soumettre la réflexion suivant:

Il est évidemment difficile de fixer exactement l'importance de la concession que l'on demandera au gouvernement autrichien, car nous désirons obtenir le maximum de ce qui est possible. Toutefois il serait peu conforme à l'importance d'une intervention directe du gouvernement français de se contenter d'un résultat, qui n'aurait pour le laboratoire Curie qu'un interêt insignifiant. Je me permets de vous envoyer, à titre de renseignement, la base que l'on pourrait adopter pour la demande.

La dernière quantité de résidu qui a été cedée au laboratoire Curie a été de <u>dix tonnes aux prix de 1000 couronnes la tonne</u>. Je pense qu'il n'est guère possible de demander moins de dix tonnes, si l'achat ne peut pas être renouvelé, et il serait désirable d'obtenir davantage.

Si cette combinaison ne pouvait pas aboutir, on pourrait demander l'assurance de pouvoir acheter régulièrement deux tonnes par an. Cette combinaison nous conviendrait également. Il serait évidemment excellent de pouvoir acheter les deux concessions a la fois.

Le gouvernement autrichien n'a pas vendu de résidu depuis trois ans, et j'estime sans pouvoir l'affirmer que la quantité de résidu accumulée peut atteindre une trentaine de tonnes. La quantité de résidu vendu pendant les années 1904 et 1905 au laboratoire Curie et à l'Académie de Vienne a, en effet, été de 20 tonnes.

En terminant, je suis heureuse de vous exprimer, Monsieur le Président, mes remerciements très sincères pour le bienveillant acceuil que vous avez bien voulu me faire, et je vous prie d'agréer l'assurance de tout mon respect

M. Curie

Professeur à la Faculté des Sciences de Paris.

À Monsieur G. Clémenceau, Président du Conseil des Ministres.

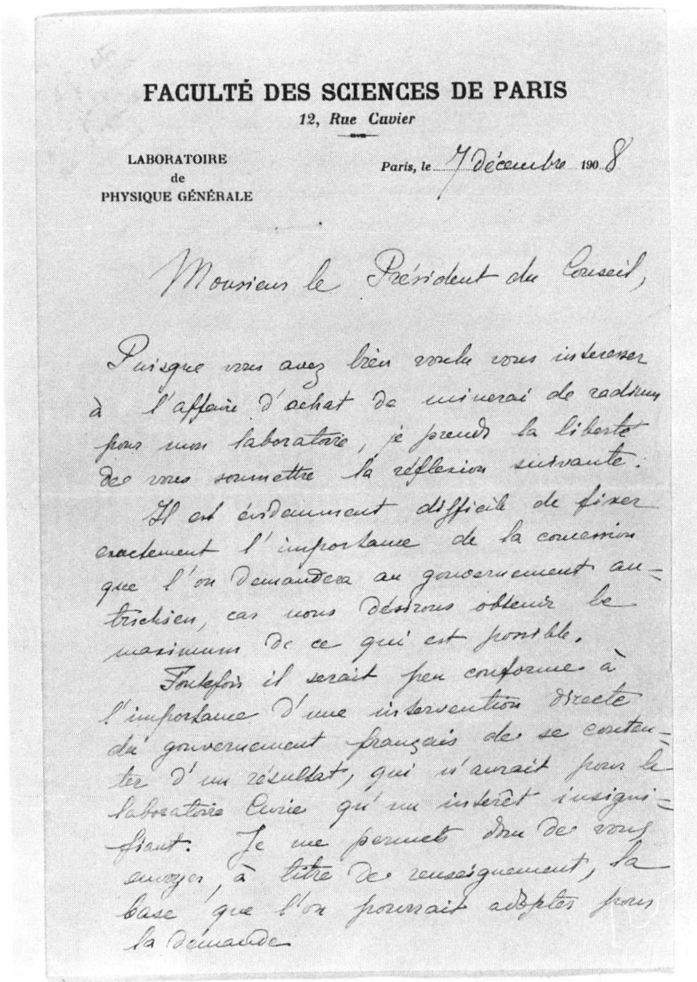

FACULTÉ DES SCIENCES DE PARIS
12, Rue Cuvier

LABORATOIRE
de
PHYSIQUE GÉNÉRALE

Paris, le 7 décembre 1908

Monsieur le Président du Conseil,

Puisque vous avez bien voulu vous interesser à l'affaire d'achat de minerai de radium pour mon laboratoire, je prends la liberté de vous soumettre la réflexion suivante.

Il est évidemment difficile de fixer exactement l'importance de la concession que l'on demandera au gouvernement autrichien, car nous désirons obtenir le maximum de ce qui est possible.

Toutefois il serait peu conforme à l'importance d'une intervention directe du gouvernement français de se contenter d'un résultat, qui n'aurait pour le laboratoire Curie qu'un intérêt insignifiant. Je me permets donc de vous envoyer, à titre de renseignement, la base que l'on pourrait adopter pour la demande

HISTORICAL NOTE

In 1898 Pierre and Marie Curie (both awarded the Nobel Prize in Physics in 1903; in 1911 M. Curie was also awarded the Nobel Prize in Chemistry) obtained the intensively radioactive element of radium. This interesting letter, in which Marie Curie asked the French Prime Minister to get from the Austrian Government permission to buy a large quantity of radium regularly, testifies to the intimate liasion between twentieth century science and politics and it shows that as early as 1908 trade in radioactive elements formed part of the diplomatic skills of Heads of State.

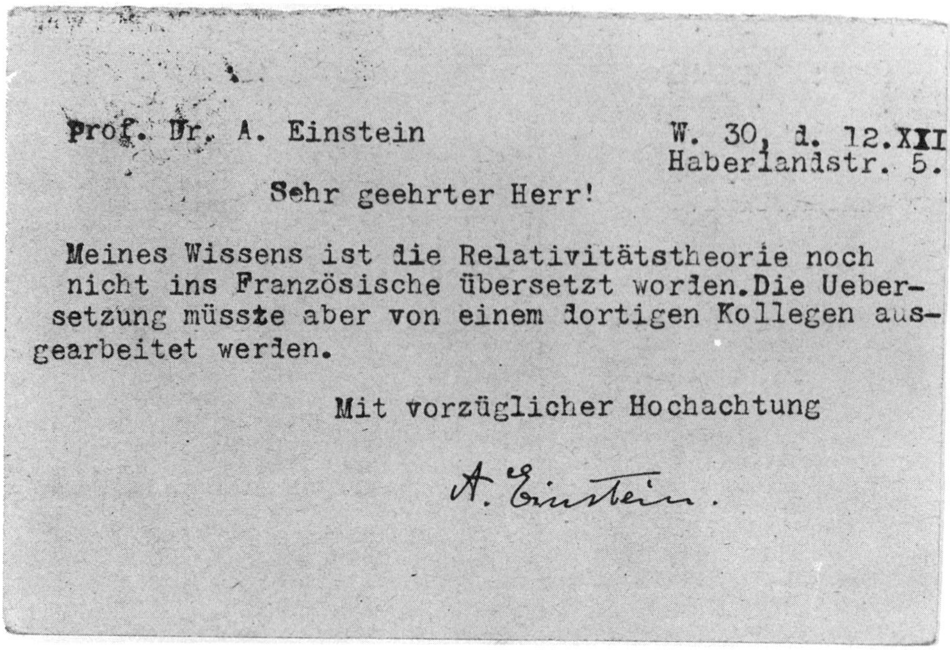

Prof. Dr. A. Einstein W. 30, d. 12.XII
Haberlandstr. 5.

Sehr geehrter Herr!

Meines Wissens ist die Relativitätstheorie noch nicht ins Französische übersetzt worden. Die Uebersetzung müsste aber von einem dortigen Kollegen ausgearbeitet werden.

Mit vorzüglicher Hochachtung

A. Einstein.

69.

EINSTEIN, ALBERT (1879-1955)

POSTCARD	German.
PLACE & DATE	Berlin, 12.12.1919.
ADDRESSEE	J. R. Herwig.
DIMENSIONS	8.9 x 13.4 cms.
SHELFMARK	Waller Delikatesskaps 1.
CONTENTS	

Einstein does not know whether his work on the theory of relativity has yet been translated into French and he wants to ensure that such a task is undertaken by a specialist. Einstein's work on the theory of relativity appeared between 1913 and 1916 and the first French translation (made by J. Rouvière with a preface by the mathematician Borel) was published in 1921 under the title *La théorie de la relativité* (Paris).

70.

[WALLER, ERIK (1875-1955)]

PORTRAIT	Oil painting.
AUTHOR	Ed. Bischoff.
DATE	1922.
DIMENSIONS	75 x 64 cms.
LOCATION	Uppsala University Library, Manuscript Room.